(POST) MODERN SCIENCE (EDUCATION)

Studies in the
Postmodern Theory of Education

Joe L. Kincheloe and Shirley R. Steinberg
General Editors

Vol. 137

PETER LANG
New York • Washington, D.C./Baltimore • Bern
Frankfurt am Main • Berlin • Brussels • Vienna • Oxford

(POST) MODERN SCIENCE (EDUCATION)

Propositions and Alternative Paths

EDITED BY
John A. Weaver, Peter Appelbaum,
and Marla Morris

PETER LANG
New York • Washington, D.C./Baltimore • Bern
Frankfurt am Main • Berlin • Brussels • Vienna • Oxford

Library of Congress Cataloging-in-Publication Data

(Post) modern science (education): propositions and alternative paths /
edited by John A. Weaver, Peter Appelbaum, and Marla Morris.
p. cm. — (Counterpoints; vol. 137)
Includes bibliographical references and index.
1. Science—Study and teaching. 2. Science—Philosophy. 3. Postmodernism.
I. Weaver, John A. II. Appelbaum, Peter Michael. III. Morris, Marla.
IV. Counterpoints (New York, N.Y.); vol. 137.
Q181 .P4945   507.1   99-053009
ISBN 0-8204-4910-5
ISSN 1058-1634

Die Deutsche Bibliothek-CIP-Einheitsaufnahme

(Post) modern science (education): propositions
and alternative paths / ed. by: John A. Weaver....
−New York; Washington, D.C./Baltimore; Bern;
Frankfurt am Main; Berlin; Brussels; Vienna; Oxford: Lang.
(Counterpoints; Vol. 137)
ISBN 0-8204-4910-5

Cover design by Dutton & Sherman Design

© 2001 Peter Lang Publishing, Inc., New York

All rights reserved.
Reprint or reproduction, even partially, in all forms such as microfilm,
xerography, microfiche, microcard, and offset strictly prohibited.

John:
To Katherine N. Hayles for her pioneering and risky work
To Don Nehlen for his ability to develop talent
To C. "We must also know how to hear and to listen."
To the People of Quakertown, Pennsylvania, and Philippi, West Virginia

In Memory of Reuben Keller (1938–2000)

Marla:
To Mary Aswell Doll and William Pinar

Peter:
To my parents Arthur (1928–1984) and Evelyn (1928–2000) Appelbaum

Table of Contents

Acknowledgments	ix
Introduction **(Post) Modern Science (Education): Propositions and Alternative Paths** John A. Weaver	1
Part One **Science as Curriculum Theory** Marla Morris	23
Chapter 1 **The Object(s) of Culture: Bruno Latour and the Relationship between Science and Culture** William E. Doll, Jr., Franc Feng, and Stephen Petrina	25
Chapter 2 **Bachelard as Constructivist** Eva Krugly-Smolska	41
Chapter 3 **The Simulacra of Science Education** David W. Blades	57
Chapter 4 **Serres Bugs the Curriculum** Marla Morris	95
Part Two **Pastiche Science: Bringing Cultural Studies of Science to Education and Education to the Cultural Studies of Science** Peter Appelbaum	111
Chapter 5 **Science Education Through Situated Knowledge** Matthew Weinstein	129

Chapter 6 — 147
Genre Analysis as a Way of Understanding Pedagogy in Mathematics Education
Susan Gerofsky

Chapter 7 — 177
A Feminist Revisioning of Infinity: Small Speculations on a Large Subject
Elaine V. Howes and Bill Rosenthal

Chapter 8 — 193
Cookbook Classrooms; Cognitive Capitulation
Dave Pushkin

Chapter 9 — 213
Modernist Traditions in Supervision and the Illusions of Science and Objectivity
Jeffrey Glanz

Part Three — 243
Pedagogies of the Cultural Studies of Science
John A. Weaver and Karen Anijar

Chapter 10 — 249
Teaching in the (*Crash*) Zone: Manifesting Cultural Studies in Science Education
Noel Gough

Chapter 11 — 275
Pedagogies of Science (In)formed by Global Perspectives: Encouraging Strong Objectivity in Classrooms
Annette Gough

Contributors — 301

Index — 305

Acknowledgments

John: I would like to thank all the contributors to this book, my family and friends for their support. More importantly, I would like to thank all of the scholars in the science studies movements. It is their work that gives us hope that science will be utilized to create a democratic world rather than a world where multinationals and modest witnesses assume it is theirs to exploit.

Marla: I wish to acknowledge the continual support Mary Aswell Doll and William Pinar give me as my academic career emerges.

Peter: My work in editing one of the sections of this book was supported in part by a grant from William Paterson University for Assigned Research Time. My chapter on Pastiche Science was nurtured by a symposium at the annual meeting of the American Educational Studies Association; I appreciate the opportunity for feedback in that setting.

We would like to also thank Lisa Dillon and her team of "eyes" for all the editing work they did to make this book more presentable.

Introduction
(Post) Modern Science (Education): Propositions and Alternative Paths
John A. Weaver

This introduction is written within the context of the science debates that have engulfed the cultural studies of science since 1993. In an indirect way this book is a challenge to critics of the cultural studies of science, who include Gerald Holton (1993) and his suggestion that the cultural studies of science is part of an antiscience movement; Paul Gross and Norm Levitt (1994) and their weak reading of the cultural studies of science; and Alan Sokal (1996) and his notion that his *Social Text* article is grounds for dismissing the whole body of work subsumed under the title, cultural studies of science. This book also constitutes an alternative path to the apologetic writings of philosophers and historians of science and their work found in Noretta Koertge's (1998) edited book *A House Built on Sand: Exposing Postmodernist Myths about Science* or Cathleen Loving's (1997) work in science education. In this introduction I want to contextualize what our purposes for this book are. On a more specific level, the individual chapters will be introduced by section introductions.

The authors in this book wish to accomplish three major goals. First, the critics of the cultural studies of science movement have particularly taken aim at poststructural French thinkers such as Bruno Latour, Jean Baudrillard, and Michel Serres. In the first part of the book, the authors offer substantive commentary on the works of selected French thinkers in order to determine what these thinkers might offer the cultural studies of science, science educators, and curriculum theorists. In this introduction I will offer some general thoughts as to why poststructuralism offers fruitful paths to understand the cultures of science.

In the second part of the book we present a vision of what a postmodern science education can/might look like. It is my position, but not necessarily that of authors to follow, that there is, at this moment, no such thing as a postmodern science education in terms of curriculum and pedagogical approaches. (I will outline in this introduction why I believe this.) The authors, therefore, offer visions as to what this might look like.

The third part attempts to address a lacuna within the cultural studies of science movement. It is my opinion that the cultural studies of science movement is the most creative movement within cultural studies, yet in terms of creating a pedagogy for the cultural studies of science, these scholars have been silent. If we expect to move beyond polemic attacks, then there has to be a movement to develop a pedagogy that informs students about the inner workings and deep impacts of the sciences on the cultures of the world.

Poststructuralism and the Cultural Studies of Science
Poststructural thought offers science education a viable alternative to modern science for numerous reasons, including its embracing of a heterogeneous, dialogic, or general economy of language. Language as medium to express and generate ideas is not seen as an efficient economy where the intentions of the author are perfectly communicated to and replicated in the minds of careful readers without ambiguity or uncertainty, nor is language seen as a means to promote any sense of closure of meaning. Instead, poststructuralism embraces the ambiguous, uncertain, and infinite. Poststructuralism abandons the modern notion embedded in science that suggests the communicative process through language can be transmitted without any noise and with perfect understanding so that good scientists can replicate what another scientist has discovered in nature. In place of a Habermasean communicative ideal, poststructuralism adopts a Bataillian general economy in which language produces through its mere utterance "excesses of energy...which by definition cannot be utilized. The excessive energy can only be lost without the slightest aim, consequently without any meaning" (Plotnitsky, 1994, p.102). To put this another way, poststructuralism adopts a stance similar to Mikhail Bakhtin's in *The Dialogic Imagination*. Bakhtin (1994) submitted that "language is not a neutral medium that passes freely and easily into the private property of the speaker's intentions; it is populated—overpopulated—with the intentions of others. Expropriating it, forcing it to submit to one's own intentions and accents, is a difficult and complicated process" (p. 294).

Critics such as Paul Gross and Norm Levitt see little value in this poststructural stance on language. In fact, Gross and Levitt object to the cultural studies of science because it embraces the whole notion of a general economy of language. Such a stance for Gross and Levitt (1994) represents an example of "muddleheadness" of the academic left which "dislikes science" (p.102). For Gross and Levitt the time is crucial for those sound, rational, level-headed thinkers to stand up for clarity because the work within the cultural studies of science is not the first time academics have tried to disguise their mere opinions as unmediated truths. Gross and Levitt (1994) warn us that "it is not without historical precedent that incoherent or simply incomprehensible opinions have had great and pernicious social effect" (p.15). Gross and Levitt object to the very ambiguity of language that poststructuralists embrace. Like their models of the world (which in their minds are not models but mirrors of nature) language has to be simple, concise, and unambiguous with unified agreement over the meaning of all crucial words, terms, and phrases within a field of study.

Fortunately, Gross and Levitt's call for an unambiguous language is impossible, as Bataille intimates with his notion of lost energy or Bakhtin with the dialogical process. Let me give an example from a philosopher of science, Thomas Kuhn, whom Gross and Levitt (1994) admire and believe to be "vulgarized and distorted" by the cultural studies of science movement, to demonstrate the general economy of language (p.56). Kuhn's book *The Structure of Scientific Revolutions* is a landmark in the history and philosophy of science, and much of the work being done today on the history, culture, and philosophy of science is (in)directly because of his efforts to expose certain issues that were once erased or ignored. Kuhn's work is as clear as one can expect, yet after other scholars began to comprehend what Kuhn was arguing there emerged a serious debate over what Kuhn meant by the term "paradigm." In response to these criticisms, Kuhn added thirty-seven pages to the second edition of his book to clarify his position on the term "paradigm." This was still not enough as books were written on the issue of Kuhnian paradigm shifts, and philosophers continue to use his term loosely and to modify the term to fit their needs.

The point here is that one word, paradigm, generated an excess of meaning that required Kuhn to write an additional thirty-seven pages to clarify his meaning of the term only to produce even more ambiguity and more attempts to clarify the term. In the name of unambiguous language, philosophers of science were unable to agree on what the term paradigm should mean. However, this ambiguity has spurred the growth of science studies from a subdiscipline that consisted of a few displaced scientists to scientists, historians, anthropologists, philosophers, cultural critics, and educators who work within multiple networks and interdisciplinary groups, publish in a wide range of journals, present at a variety of conferences, and speak numerous, and often incommensurable, languages that shape the world they see and interpret. It was a general economy of language that created this infrastructure of science studies, and as long as the language of science studies remains a general economy that infrastructure should only continue to grow and flourish.

While Gross and Levitt may cringe at the thought of a general economy of language, to many it is the source of creativity, imagination, quality scholarship, and interesting debates. It is also one of the best guarantees we have against authoritarianism in science and society: a general economy of language prevents a dictatorship of meaning. I want to suggest that this general economy of language is one aspect of the cultural studies of science that irks Gross and Levitt. As Babette Babich mentions, one of the premises underlining not only Gross and Levitt's book but the other recent books and articles as well is a belief that if only unambiguousness and conciseness (or sanity) can be restored to science then so will the undisputed and unquestioned position of science in society. "Nip the obscurantism, bad writing, and thought patterns of deconstruction and postmodernism," Babich (1997) suggests, "in the bud, and funding levels for the supercollider will be up to speed in no time. No more social constructivist talk: Bring reality back" (p.25). In other words, if those left wing academics, with their books and articles on the lack of citizen representation in science policy and the lack of certainty in science, would stop poking their heads into places where they do not belong, then those right wing politicians on Capitol Hill, who apparently spend hours reading this scholarship, would restore science to its proper place on top of the hierarchical structure of American society.

The power of unambiguousness is also found in its ability to control what are legitimate and proper ways of thinking. Patti Lather (1996) adds to this line of thinking when she writes: "Clear speech is part of a discursive system, a network of power that has material effects. Premised on incorporating a particular form of everydayness into public statements as tools of circulation and naturalization, charges of 'not in the real world' or 'too academic' have particular kinds of effects" (p.528). In regards to science these particular kinds of effects include what forms of research are sanctioned as legitimate, what interpretive parameters are deemed acceptable, and who is labeled rational and logical. The struggle over (un)ambiguous language is a struggle over who can exist in science, who can decide the meaning of science, and who counts as a legitimate voice in charting the future of science and society. This warning against unambiguous language in no way denies the importance of science in our lives; it merely warns against those who wish to eliminate or limit the role of nonscientists in significant scientific decisions that influence all sectors of society. The cultural studies of science not only offer insights into how science works but also how it can work for a democratic society. The call by cultural studies of science scholars for the end of the myth of unambiguous language is an encouraging beginning.

The strength of poststructuralism is also found in its critique of what Sharon Traweek and Donna Haraway calls "modest witnessing." Based in and on the traditions of Robert Boyle and Francis Bacon, modest witnessing is the culture of no culture, in which the scientist is a mere insignificant conduit through which nature speaks. The modest witness is not unlike a missionary willing to proclaim his purity and sanctity to the world and convert all to his calling. As a modest witness he "is objective; he guarantees the clarity and purity of objects. His subjectivity is his objectivity. His narratives have magical power" (Haraway, 1997, p.24). When it is time to proclaim what he believes, the modest witness adopts a "naked way of writing, unadorned, factual, compelling," to associate his proclamation with a higher power. It is through this "naked way of writing" that the modest witness can convince himself and others that what he is saying, interpreting, and describing is unmediated by his own subjectivity and humanness since these traits are not pure and reliable. These traits have to be denied faithfully for the modest witness to be the prophet of all that is true and natural.

Critics of the cultural studies of science have argued that the lampooning of the modest witness is a form of relativism that denies the existence of an independent reality. Sokal (1996) writes: "There is a real world; its properties are not merely social constructions; facts and evidence do matter. What sane person would contend otherwise? And yet, much contemporary academic theorizing consists precisely of attempts to blur these obvious truths" (p.63). In his coauthored work with Jean Bricmont (1998), Sokal adds: "I'm a stodgy old scientist who believes, naively, that there exists an external world, that there exist objective truths about that world, and that my job is to discover some of them"(p.269). Sokal certainly would have a legitimate criticism if such a stance actually were taken by anyone in the cultural studies of science. However, the fact of the matter is that reality is not being contested nor is the existence of an external world free from any human biases. What is being contested is the notion that anyone, whether it is the "good" scientist or the reality-denying cultural critic, can know reality or its external truths without some form of mediation, whether it is language or a laboratory instrument. What is being contested is the whole notion that started with Boyle and Bacon that we can know the world in its uninterrupted splendor. The cultural studies of science, through its historical case studies, philosophical readings, anthropological studies, and linguistic formations suggest we could never know the world in an unmediated fashion, and we should learn to embrace science in spite of all of its human biases and flaws because it is the best human creation invented to this point to understand/create/invent nature.

From this perspective, poststructuralism is a theoretical approach that simultaneously critiques the notion of the modest witness and offers a critical reading of how science is constructed and used sometimes to justify certain policies, eliminate unwanted voices, and apologize for all its flaws and humaness in order to shroud science behind a cloth of god-like importance and ability. Moreover, poststructuralism presents an alternative reading of science that, taking from Haraway's (1991) "Cyborg Manifesto," is "committed to partiality, irony, intimacy, and perversity" and is "wary of holism, but needy for connection" (p.151). Poststructuralism is not committed to these notions because they are trendy, postmodern buzzwords but rather because they represent the human condition and the impossibility of transcending the human tendency to be partial, ironic, intimate, perverse, incomplete, and uncertain.

Poststructuralism offers an alternative to the common accusation that the cultural studies of science scholars are relativists. The charge of relativism ignores the influence Foucault, Derrida, Latour, and others have had on these scholars. When one labels a scholar with whom one disagrees a "relativist," it not only announces a discursive power play based in the authority of modest witnessing but also erases the role power holds in the thinking of cultural studies of science scholars. Relativism is a simplistic modernist concept that assumes away power in the opinion making process. Relativism claims no one person holds a direct and unmediated link to a transcendent authority that knows what truth is, but it completely ignores how those with power construct and sanction legitimate ways of thinking, thereby denying the flourishing of an infinite amount of ways of thinking. The cultural studies of science movement is interested not in proclaiming the legitimacy of all opinions, but rather it is interested in the very power dimensions that relativists ignore. It is power which decides what will be called science, who will be called a scientist, and what ways of thinking will be deemed scientific. It is because of power that strict methodological and epistemological boundaries exist as to what is truth and what is legitimate knowledge. Those who question this power risk incurring the wrath of those who embody this power because as a dissenter they are banned from reaping the benefits of the scientific community. Dissenters are discursively constructed in society as unreliable, invalid, and dangerous. The only way relativism enters into the discussions within the cultural studies of science movement is when traditional modest witnesses such as Sokal, Bricmont, and Holton utilize it to limit the influence of the cultural studies of science.

By utilizing the term "relativist," Sokal, Bricmont, and others ignore another dimension embedded in the concept of relativism that is in direct conflict with a principle underpinning the cultural studies of science. As Babette Babich (1994) points out:

> The relativistic position takes all philosophical viewpoints...as ultimately equivalent, that is, equal one to another as well as to its own position. This conception equates all perspectives and presupposes an absolute, conceptually essential, if admittedly or at least decidedly unattainable, standpoint. The unattainable absolute is the imaginary Archimedean standpoint from which all perspectives are relatively equivalent. Axiomatic for any relativism, this mediating absolute permits the relativist to claim the same right or value for all philosophic positions (p.49).

That is, relativists posit an axiomatic position that claims they are able to judge all possible interpretations equal. From this Archimedean position the relativist claims the power to speak for all possible interpretations, thereby erasing all alternative voices, including those who claim the existence of an absolute, unmediated position and those, like cultural studies of science scholars, who contest such an absolutist position. Cultural studies of science object to the relativist position not only because the Archimedean claim to judge all positions equal ignores power as mentioned above, but also because it privileges the epistemological foundation of relativism. The cultural studies of science movement, among other things, is in part a manifestation of the postmodern movement that places in doubt and undermines all epistemological foundations, including those that undergird traditional science, relativism, and the cultural studies of science. Such a stance does not deny the existence or discourage the creation of epistemological foundations. What it does attempt to do is problematize all epistemological foundations in general and absolutist foundations in particular. Sokal, Bricmont, and others ignore the antiepistemological project of the cultural studies of science and as a result conveniently label the movement relativistic.

Let me provide one example from Sokal and Bricmont's work *Fashionable Nonsense* in order to demonstrate how they play discursive power games to label the cultural studies of science movement as relativist in order to dismiss any insights it might provide. Their favorite target when constructing their "postmodern" relativism argument is Bruno Latour. They purport that "Latour's writings provide considerable grist for the mill of contemporary relativism..."(1998, p.9) and such relativist thinking "(as well as other postmodern ideas) has effects on the culture in general and on people's ways of thinking" (1998, p.99). They cite Latour's work on Einstein's relativity theory and one of his early books *Science in Action*. Like their critiques of other French intellectuals they offer no substantial analysis of Latour's work, nor do they provide a context from which they draw quotes. Instead, they throw out lengthy quotes from some of his work and proclaim Latour is a symbol of relativism. In their perusal of Latour's work they conveniently have omitted one of his latest works, *We Have Never Been Modern* (1993). This work goes to great lengths to dismiss all forms of relativism. For Latour (1993), relativism is a pernicious form of cultural imperialism in which "Relativists, who strive to put all cultures on an equal footing by viewing all of them as equally arbitrary codings of a natural world whose production is unexplained, do not succeed in respecting the efforts collectives make to dominate one another" (p.108). In other words, attempts to construct a relativist world view ignore and erase the power plays that are taking shape as different collectives or cultures converge to control others either through political, cultural, or discursive powers. Latour rejects relativism on the grounds it dismisses the role of power to define reality. In their zeal to dismiss French intellectual thought, Sokal and Bricmont draw selectively from Latour's work and construct the only discursive image they can from their absolute realism: relativism. For Sokal and Bricmont, Latour and other cultural studies of science scholars cannot be searching for alternatives to modern thought because in the dichotomous, either/or thinking of Sokal and Bricmont, the only alternative to rational thinking is relativism. There can be no other forms of rational thinking; any attempt to search for alternative forms is just "fashionable nonsense."

(Post) Modern Science (Education)

It is my contention that there is no such thing as a postmodern science, but the possibilities for a postmodern science education curriculum are not only possible but necessary. If we take postmodernity to imply a sense of indeterminacy, a movement away from closure; a distrust for universal, generalizable, united theories of nature and society; the ambiguity of language; and antiessentialism and antifoundationalism, then clearly there is little possibility that science has much to offer the postmodern movement except for some heuristic metaphors. Many scientists may have changed their metaphors from clocks and symmetry to bees, ants, and asymmetry, but their vision and goal remain the same: naming that which they study as universal while ignoring the naming process. Science, even chaos theory, is exclusively empirical (that which can be labeled observable or a matter of fact "discovered" in the laboratory and can be quantified and falsified while denying that any metaphysical assumptions are often created to construct models and thought experiments about the world), with a high disregard for, or at best a high level of embarrassment toward, experience, intuition, and hunches. Science is also methodologically conservative in the sense that nonexperimental methods are either not valued or just incompatible with the goals of traditional science. Moreover, science, with a few exceptions such as Humberto Maturana and Francisco Varela (1987) and Ilya Prigogine (1996), is reductionary and deterministic even when it comes to such admittedly complex matters as neural networks and dynamic systems. Postmodern science education has given up what Maturana (1987) calls the "temptation of certainty," but I am skeptical whether science will ever give up this illusory quest (p.18).

When I say certainty is an illusion I am not implying we should not attempt to understand nature, model natural phenomena, or interpret that which we "discover" or think we see. The satisfaction of science is not discovering truths or universals but is found in the creative process of inventing models and interpretive approaches that provide answers to our problems without forgetting that we are using models and interpretations. When we search for truths and universals we live the illusion that we can capture all that there is to know in the universe just as Stephen Hawking believes we can know all that we need to know about the universe. We live the illusion that knowledge is static and by our very inquisitiveness and probing we do not create more knowledge and possibilities. In this sense, our new metaphors should not be ants, event horizons, or bees but textual metaphors such as the universe as Scholes' textuality (1998) or Latour's Factishes (1999). John Holland is one scientist who seems to realize the textual dimension of science at least when it comes to modeling. In his work on complex adaptive systems Holland (1995) reminds his readers that:

> Modeling, it should be clear, is an art form. It depends on the experience and taste of the modeler. In this it is much like cartooning, especially political cartooning. The modeler (cartoonist) must decide which features to make salient (exaggerate), and which features to eliminate (avoid), in order to answer the questions (make the political point) (p.11).

Does Holland's splendid and refreshing honesty discredit his important work in understanding immune systems, economies, nervous systems, and ecosystems? I don't think so. In fact, to me, his self-reflection makes his work more enticing and more worthy of public and monetary support because he is not making monopolistic claims that his path is the only path to understanding immune systems and so on. His recognition of the literary and political dimension of science encourages others to develop their own paths in finding cures to diseases such as AIDS or finding better understandings of the fluctuations of the stock market. Holland's recognition that modeling is a human endeavor, not a mirror of nature, accepts that science, like all forms of knowledge, is a general economy of meaning in which some dimensions will always escape interpretation and the model created will have only limited success in creating understanding of a natural phenomenon.

Holland, I believe, is a rare exception in science, at least in regards to his views on modeling. This unwillingness amongst scientists to see the human hand inextricably tied to the acts of science, I think, is one reason why it is a fruitless task to argue that there is a postmodern science that can exist or that such a science is on the horizon. However, it is highly beneficial for educators to begin the process of thinking what a postmodern science education might look like. Before I get into the benefits of developing a postmodern science curriculum I want to explain more about my skepticism concerning the notion of a postmodern science. To do this I want to utilize examples from the writings of Niels Bohr and Werner Heisenberg and recent scientific developments in chaos theory.

In regards to Niels Bohr and Werner Heisenberg, it is customary to cite passages from their writings to present philosophical evidence against the idea of pure objectivity, unambiguous language, and a noncontradictory reality. For instance, Werner Heisenberg's philosophical essays contain constant reminders to his readers that no matter what experimental device or observation tool we utilize to see the world, it influences how we see the world. In other words, there is no possible way to see the world unmediated. As Heisenberg (1958) noted in *Physics and Philosophy,*

> Natural science does not simply describe and explain nature; it is part of the interplay between nature and ourselves; it describes nature as exposed to our method of questioning. This was a possibility of which Descartes could not have thought, but it makes the sharp separation between the world and the I impossible (p.81).

However, Heisenberg (1952), in a collection of his earlier philosophical essays titled *Philosophical Problems of Quantum Physics*, warns against extending the impact of quantum physics too far beyond its original field of study. He notes:

> it is equally wrong to speak today of a revolution in physics. Modern physics has changed nothing in the great classical disciplines of, for instance, mechanics, optics, and heat. Only the conception of hitherto unexplored regions... (p.18).

I would add to Heisenberg's list the methodological approach utilized in science. In spite of Heisenberg and Bohr's growing awareness that the tools one utilizes to see the natural world construct in part that world they did not seek to open the doors of science to alternative methodological approaches and probably were right in not doing so. Moreover, Heisenberg's philosophical arguments against nonmediation do not question the epistemological foundation of science in which a Hegelian hierarchy is instituted to rank order knowledge, starting with empirical knowledge at the top, with any forms of knowledge that contain the least bit of intuitive or personal knowledge (e.g., hunches and storytelling) relegated to the bottom. Postmodernity, unlike traditional enlightened science, values personal knowledge as much as empirical knowledge and neither places sources of knowledge into an arbitrary hierarchy nor accepts the universal certainty of empirical knowledge. Heisenberg clearly would object to such notions of knowledge.

Bohr is often cited by postmodernists but is an uncertain ally as well. In certain circumstances, Bohr recognizes the ambiguousness of language in describing a natural and human phenomenon. For instance, Bohr (1987) believed that, given the limitations of our ability to measure simultaneously the momentum and position of a particle, it was impossible to present "an unambiguous definition of the state of the system" (p.54). Our language was limited by what we could see and measure; therefore, there was always an indeterminate dimension about what we could see and how we could describe what we saw. Even classical concepts are rendered ambiguous in the subatomic world of quantum physics. As he notes:

> We must bear in mind that the possibility of an unambiguous use of these fundamental concepts solely depends upon the selfconsistency of the classical theories from which they are derived and that, therefore, the limits imposed upon the application of these concepts are naturally determined by the extent to

> which we may, in our account of the phenomena, disregard the element which is foreign to classical theories and symbolized by the quantum of action (p.16)

In other words, within the classical model of physics the language which emerged to describe what physicists wished to explain is unambiguous within that context. However, once physicists attempt to apply this language to describe what they see in the quantum world, language loses its unambiguousness.

It is in Bohr's concept of complementarity that we find simultaneously his radical contribution to science and his commitment to traditional enlightened science. Bohr's notion of complementarity can be seen as the simultaneous existence of two realities, two modes of interpretation within one world that are mutually exclusive and complete as much as possible within their own realm. As Plotnitsky (1994) points out, "Bohr always maintained the relative indispensability of classical concepts and descriptions, for both historical and theoretical reasons; and he conceives of this type of closure in very broad terms, well beyond the domain of physics or natural science" (p.130). In his own words, Bohr expresses his commitment to modernist principles and modernist science:

> It would be a misconception to believe that the difficulties of the atomic theory may be evaded by eventually replacing the concepts of classical physics by new conceptual forms. Indeed,...the recognition of our forms of perception by no means implies that we can dispense with our customary ideas or their direct verbal expressions when reducing our sense impressions to order (Plotnitsky 1994, p.130).

At best, postmodernity and Bohr have an uncomfortable relationship. Where Bohr recognized the limitations of certainty at the quantum level, he reinforces certainty at the macro level of the world. He re-establishes an "imaginary order," which Robin Usher and Richard Edwards (1994) suggest is privileged by empirical science and its elimination of any form of knowledge that is not based in "observation and experimental phenomena" and "illusory wholeness"(p.73).

I am not suggesting that postmodernity is against empirical science. What I am suggesting is that if we are looking to science to end the modernist quest for "illusory wholeness" that invites closure, universal laws that erase differences, and reductionist thinking that simplifies the irreducible world of human relations, we will not find much evidence to support our quest. I think our desire to read postmodern phenomena into science stems from the modern hierarchies of knowledge that placed the sciences above nonscientific forms of human knowledge. We are still looking for legitimacy and approval from the sciences when we should be working in complementarity with, not in competition with, the sciences to chart our own course of action since our goals are incommensurable to the epistemological foundations of modern science. In this sense Bohr's notion of complementarity has much to offer us as a guidepost to determine how we should approach our goal of a postmodern science education. Again I refer to Plotnitsky's (1994) work:

> [What] the theory of complementarity offers is, however, not incomplete. On the contrary, it is a complete theory of its data,...Complementarity is as complete as a theory can be under the available conditions and constraints, such as those of incompleteness that make it impossible to have a complete picture in the classical sense (p.128).

Chaos theory, like Bohr and Heisenberg, is often called postmodern, and there are concepts within the theory that support such a claim. However, there are also strands of thought within chaos that have as their core premise the reinstitution of a unified theory, or at the very least, a reductionist notion that a few simple laws guide even the most complex systems. As Hayles (1991) points out, "chaos theory has a double edge....On the one hand, chaos theory implies the Newtonian mechanism is much more limited in its applicability than Laplace supposed. On the other hand, it aims to tame the unruliness of turbulence by bringing it within the scope of mathematical modeling and scientific theory" (p.15).

The concept of initial conditions in chaos theory is part of that double edge. In chaos theory, two systems can begin with near precise initial conditions but, given enough time and perturbations, spiral out of control to the point the two systems end up at two points that are dramatically different and unpredictable. Such an unpredictable outcome renders the knowledge of the initial conditions or the system's past meaningless. Stephen Kellert (1993) discusses the impact of this phenomena on the philosophy of science: "Chaotic dynamics will take the tiny indeterminacies...and stretch them into huge variations, dilating the smallest patch until, as some sufficiently distant time in the future, almost anything is possible" (p.73). From this realization, Kellert concludes that "determinism" or predictability "is not so much proven false as rendered meaningless" (p.74).The importance of initial conditions in dynamic systems can also be seen in traditional Newtonian models as well. Ian Percival suggests that even in the swings of pendulum, predictability is impossible in the long term. Percival (1991) notes:

> When the bob moves slowly near to a point midway between the magnets, it is affected almost equally by the force from each magnet. Its future notion becomes extremely sensitive to small changes in its present position and velocity, so the motion is chaotic....To make the same prediction after four swings, its position would have to be measured to within the size of a bacterium, and after nine swings, to within less than the size of an atom. The pendulum obeys Newton's deterministic laws, but any attempt to predict its future behaviour over long times will be defeated (p.13).

The classical principles of science such as generalizability, the unity of knowledge, universal laws, and predictability, seem to be placed in doubt while support for a postmodern world view of difference, indeterminacy, and local knowledge appears promising.

However, as Katherine Hayles points out in the midst of these unsettling discoveries, there has emerged a reaffirmation of reductionist thinking and, possibly in the future, predictability. As an example of this desire to restore reductionist thinking and predictability to the lexicon of science, Hayles cites Mitchell Feigenbaum's work in nonlinear systems. Feigenbaum recognizes that a basic premise of chaos theory is uncertainty when he states "'unless the initial conditions are known to *infinite precision*, all known knowledge is eroded rapidly to future ignorance.'" (Hayles, 1991, p.153). However, as Hayles notes, Feigenbaum holds out hope for the return of traditional, enlightened science when he also discovered in his work that there are still some universal tendencies within chaos theory including the discovery that all the nonlinear systems he studied "approached chaos at the same rate" (Hayles 1991 p.153). This has encouraged some chaos theorists to generalize that there may be some aspects of the natural world that a Newtonian system does not explain and that the world is more complex than classical theories are able to explain, but the simple, orderly, universal laws of nature are just embedded deeper into the structure. For instance, these proclaimations by Feigenbaum and others have prompted Ian Stewart and Jack Cohen (1997), in their quest to understand evolution, to proclaim that "what we really need...is Simplicity Theory, an effective and relatively painless way to extract the big simplicities from the underlying rules" (p.46). Even though chaos theory has introduced uncertainties, unpredictability, and complexity, scientists have worked hard to tame, once again, the indeterminancies of life.

Postmodern science education should not try to ignore the tendency of scientists to seek out universal laws, assert generalizations, and reduce the complex to the simplistic. As Stewart and Cohen (1997) point out, the reductionist approach may be a nightmare, but it still has produced enough evidence to support its continuance as a model to explain some natural phenomena (pp.46–47). What a postmodern science education should do is discuss how scientists construct models to understand the world, and these models may produce evidence to support their usage, but they are only models or simulations of the world, not correspondent representations. A postmodern science education can critique the myth of a correspondence theory of truth without condemning scientists for constructing models. The only thing scientists should be criticized for is believing their own illusion that the models they construct really, somehow, are representations of the natural world rather than representations of their minds. In order to create an understanding of representations of the mind, postmodern science education will need to shift its focus from metaphors that permit scientists to ignore the power of representations of the mind to construct nature to literary metaphors that take into consideration how scientists represent nature, such as Bèrubè's (1994) political model of representation, Landow's (1997) hypertext model of representation, or Bahktin's (1994) dialogical model of representation.

A postmodern science education consists of other dimensions besides the discussion of models of representation. A postmodern science education needs to discuss the role power plays in the making of science. There are two crucial types of power that need to be discussed not only in regards to science but all dimensions of human knowledge: 1) normative power and 2) monetary power. The first type of power requires us to re-think popular notions about how science gets done. In the spirit of the modest witness, the assumption is often made that universal laws would have been discovered around the time they were even if the discoverer never existed. This approach assumes that natural laws speak through individuals while the creative, interpretive, and imaginative abilities of individuals to force nature to speak according to their models and theories are subsequently insignificant. According to this approach to science the individual scientist is simultaneously silent and a genius. He is able to get natural laws to reveal themselves without saying a word that was not one of nature's own words in the first place.

Normative power is found in who is allowed to speak as well. This takes the form of who gets credit for the discovery, who is given the necessary credentials to speak, and who is permitted into the community in the first place. The rhetorician of science, Alan Gross, reveals that the power to claim a discovery is often a political act. In writing about the disputes between Newton and Leibniz over the "true" inventor of calculus, Gross (1989) explains how the president of the Royal Society commissioned "an international committee to investigate the twenty-year controversy between Newton and Leibniz concerning priority in the invention of the calculus" (p.95). The committee ruled in Newton's favor, which is well known, but what is not mentioned or discussed is that Newton was the president of the Royal Society at the time and the author of the report that concluded he invented calculus.

A postmodern science education should not just discuss ways in which normative power influences committee reports but also how it determines who should be promoted, who should be appointed to influential boards, and who should serve as distinguished chairs. Students need to know that science does not work in a vacuum but functions within a cultural and historical context that shapes how people think and act. They also need to know how those boundaries of thinking and acting are patrolled, sometimes ruthlessly to the point where certain people are denied access to the structural support that is needed to do science successfully. In regards to monetary power, a postmodern science education has to inform students of the ways in which money influences scientific decisions. Students need to be made aware of how funding decisions are made by governmental boards, such as the National Science Foundation, and nongovernmental boards, such as philanthropic boards. In this regard, they need to be made aware that decisions are often made on ideological grounds, not on utilitarian or meritorious grounds. For instance, Ellen Messer-Davidow (1995) reveals there is a rapidly growing network of conservative think tanks (e.g., Heritage Foundation, American Enterprise Institute, Hudson Institute, etc.) and conservative foundations (e.g., the Olin, Bradley, Scaife, Coors, etc.) which will not fund research that does not pass an ideological litmus test. Students need to be aware that such ideological grounding of research funding is not new but is historically based. Dorothy Ross (1991) points out, for instance, that the Ford, Carnegie, and Rockefeller foundations in the 1920s funded research that was not necessarily meritorious but provided insights into how to better control and manage an uneducated and potentially unruly population in America.

In terms of monetary power, a postmodern science education has to deal with the growing influence of multinational corporations on the scientific process. While foundations support funding based on ideology, multinational corporations are funding laboratories and endowing chairs so university scientists can create profitable products for the companies. The growing support from multinational corporations raises serious doubts about the traditional autonomy of the scientist. If support from foundations did not burst the myth of the apolitical scientist, then certainly the alliance between science and multinational corporations will. The disturbing part of the apolitical myth in science is that when it comes to receiving funds from foundations or multinationals, the myth is suspended and a nonissue, but when it comes time for corporations and scientists to take responsibility for an ill-conceived or unethical product, the apolitical myth is raised as a banner which is designed to protect scientists from any liability for their complicity. A postmodern science education must contest the growth of multinational corporation science because as Andrew Ross (1996) warns, "in today's marketplace, it is the corporate bottom line, expressed, for example in the golden rule of patent protection that decides when public sharing of scientific information ends and when product development and monopoly control over profits takes over" (p.179).

For no other reason, a postmodern science education should include a discussion of power and the other issues discussed above because the primary goal is the construction of a democratic approach to science. Postmodern science education needs to put an end to the notion that the public is too ignorant to understand and take part in major scientific policy decisions. As Donna Haraway (1997) reminds us, "the 'public,'" has been "conceptualized as a passive entity with 'attitudes' or 'understandings' but not as a bumptious technoscientific actor" (p.94). A postmodern science education needs to demand the public be treated as a "bumptious" actor in all aspects of society that impinges upon the quality of their life. At its heart, a postmodern science education is not only about the teaching of science but the complexity of knowledge (in spite of simplistic scientific models) and life and the importance of making sound public policy decisions that are not done for ideological or profit motives but the sustaining of a quality life. This is why a postmodern science education should pay more attention to the education component and less trying to discover dimensions of science that can support a postmodern perspective.

Pedagogies of the Cultural Studies of Science
T. Hugh Crawford (1997), in his contribution to the science debates, remarks that according to the work by the historian of science, Steven Shapin, an unmediated perspective of nature is a matter of "teaching

practitioners how to see, what to look for, and what conclusions to draw" (p.56). In other words, traditional science is more than doing science but about teaching science a certain way. For traditional enlightened science it is a matter of teaching practitioners how to construct nature, erase themselves, and to rationalize this pedagogy as an accurate representation. If the cultural studies of science seek to become a viable alternative to traditional enlightened science, it will have to begin to invent a pedagogy that offers alternative ways of seeing nature, science, and the world.

Along similar lines, Thomas Kuhn believed that normal science is about pedagogy. "Normal science," Kuhn (1970) concluded, "is predicated on the assumption that the scientific community knows what the world is like" (p.5). And the way in which most scientists came to believe in this assumption was through a pedagogy driven and constructed by textbooks within the fields of science. Kuhn writes that inevitably much of what took place during a scientific revolution is lost. The struggles of breaking an old paradigm, the competition over how the new information will be interpreted, the errors and unsuccessful replacements for the old paradigm, and, certainly, most, if not all, of the politics of a new paradigmatic revolution, are erased. Textbooks, Kuhn suggests, in order to

> fulfill their function...need not provide authentic information about the way in which those bases were first recognized and then embraced by the profession. In the case of textbooks, at least, there are even good reasons why, in these matters, they should be systematically misleading....Textbooks...being pedagogic vehicles for the perpetuation of normal science, have to be rewritten in whole or in part whenever the language, problem-structure, or standards of normal science change and once rewritten, they inevitably disguise not only the role but the very existence of the revolutions that produced them....Textbooks thus begin by truncating the scientists' sense of his discipline's history and then proceed to supply a substitute for what they have eliminated (p.137).

In traditional enlightened science this elimination process has come to mean the perpetuation not only of a process of unmediated discovery of natural laws but also a sense that this process of discovery is part of a continuity in which politics, alternative interpretations, misinterpretations, and the overlooking of earlier discoveries that support a later paradigm shift (e.g., Galileo's work in astronomy and Mendel's work in genetics) are nonexistent factors in the scientific practice. This elimination process is what Bruno Latour (1987) has called "ready made science."

Katherine Hayles in her work on chaos theory provides an example of how this elimination process Kuhn and Latour outline is not only still in effect but is quite efficient and expedient. In describing the process in which the discovery of fractal geometry came about, Hayles notes a shift in how Benoit Mandelbrot rewrites the past in his telling just after the discovery and a few years later when it was now known that what Mandelbrot discovered was indeed an important event in mathematics and science. Writing about the difference between the two editions of Mandelbrot's *Fractal Geometry*, Hayles (1991) reveals:

> In the first edition, when fractal geometry was a new idea, Mandelbrot took pains to present it as a continuation and extension of earlier ideas. By the second edition, when fractal geometry had become a recognized field that was attracting more researchers daily, Mandelbrot rewrote key passages to emphasize the uniqueness and singularity of his vision. There is an obvious political agenda to this jostling for position; it is no secret that Mandelbrot is a contender for a Nobel Prize (p.168).

History is a process of rewriting, reinterpreting, and reinventing the past. The problem with traditional enlightened science is that it erases this process of rewriting, reinterpreting, and reinventing as it constructs the illusion of unfettered progress and nonmediated discoveries of truth. What we should be seeking to create is a pedagogy that offers us explanations of what Latour (1987) refers to as "science in the making."

The way of seeing what Crawford mentions, the way of remembering science's past(s) that Kuhn and Latour mention as "ready made science," and the way of rewriting science that Hayles demonstrates, have been successful pedagogical strategies for traditional enlightened science to continue its dominance as the most reliable and realistic source of knowledge we have of the world and society. Within this pedagogy is a process of erasure in which viable alternatives are dismissed either as a part of the unmentioned past or as a part of what is seen as the dangerous allure of irrational, romantic desires that traditional enlightened science is constantly battling.

The cultural studies of science movement, however, has failed to challenge the pedagogical style of traditional enlightened science. The cultural studies of science movement has offered an anthropological, philosophical, historical, sociological, political, and cultural interpretation of traditional enlightened science but has ignored the pedagogical. The third part of this book attempts to begin the process of creating what we call pedagogies of the cultural studies of science.

With pedagogies of the cultural studies of science, we can begin the process of getting our students to see the natural world not as a nonmediated realm of reality, but as a human endeavor in which our role is crucial in constructing science and naming nature. These new pedagogies will not only challenge the traditional enlightened pedagogy of nonmediation, but they will also infuse a much–needed self-reflective dimension into the scientific process. No longer would it be assumed that science will automatically progress for the betterment of society or automatically create something that is for the public good. Rather these new pedagogies of the cultural studies of science would ask how recent discoveries might affect society and what public policies represent the best usage of these discoveries. These pedagogies of the cultural studies of science would ask not only how do I construct myself within the processes of science but how do I construct others. Do I construct myself and others as an active or an invisible agent within science?

Moreover, these pedagogies of the cultural studies of science would be about creating alternatives not only to traditional enlightened science but to any paradigm that emerges as a dominant way of seeing the world and attempts to reify the world as static. These pedagogies are about asking important questions about learning. We should follow Sharon Traweek's (1996) lead and ask ourselves: "Why should there be only one way to think well, only one way to have fun with our minds? Why is mental monogamy required?...Does thinking without singularities mean we cannot think carefully about ourselves, other human beings, and our phenomenal world?" (p.148). Such questions get at the heart of creating alternative pedagogies that do not erase or condemn such alternatives as irrational and illogical. As soon as we can create alternative pedagogies that break the grip of those stifling binaries that infest the minds of traditional enlightened science, we can create dynamic classrooms that turn science into a process of discovery and intellectual debate rather than a process of replication and rote memorization.

References

Babich, B. (1997). The hermeneutics hoax: On the mismatch of physics and cultural criticism. In *Common Knowledge* (6:2) 23–33.

Babich, B. (1994) *Nietzsche's philosophy of science*. Albany: SUNY.

Bakhtin, M. (1994) *The dialogic imagination*. Austin: University of Texas.

Bèrubè, M. (1994) *Public access: literary theory and American cultural politics*. London:Verso.

Bohr, N. (1987). *The philosophical writings of Niels Bohr*, Vol. 1. Woodbridge, CT: Ox Bow.

Crawford, T. H. (1997). Screening science: pedagogy and practice in Williams Dieterle's film biographies of science. In *Common Knowledge* (6:2):52–68.

Gross, A. (1989) The rhetorical invention of scientific invention: The emergence and transformation of a social norm. In *Rhetoric in the human sciences: Inquiries in social construction*, edited by Herbert W. Simons, 89–107. London: Sage.

Gross, P., & Levitt, N. (1994). *Higher superstition: The academic left and its quarrels with science*. Baltimore: Johns Hopkins.

Haraway, D. (1991). *Simians, cyborgs, and women: The reinvention of nature*. New York: Routledge.

Haraway, D. (1997). *Modest_witness @Second_Millennium.FemaleMan © meets OncoMouse ™*. New York: Routledge.

Hayles, K. (1991). *Chaos bound: Orderly disorder in contemporary literature and science*. Ithaca: Cornell.

Heisenberg, W. (1952). *Philosophical problems of quantum physics*. Woodbridge, CT: Ox Bow.

Heisenberg, W. (1958). *Physics and philosophy: The revolution in modern science*. New York: Harper.

Holland, J. (1995). *Hidden order: How adaptation builds complexity*. New York: Addison and Wesley.

Holton, G. (1993). *Science and anti-science*. Cambridge: Harvard.

Kellert, S. (1993). *In the wake of chaos*. Chicago: University of Chicago.

Koertge, N. (Ed.) (1998). *A house built on sand: Exposing postmodernist myths about science*. New York: Oxford.

Kuhn, T. (1970). *The structure of scientific revolutions*. 2nd ed. Chicago: University of Chicago.

Landow, G. (1997). *Hypertext 2.0: The convergence of contemporary critical theory and technology*. Baltimore: Johns Hopkins.

Lather, P. (1996) Troubling clarity: The politics of accessible language. In *Harvard Educational Review* (66:3): 525–545.

Latour, B. (1987). *Science in action: How to follow scientists and engineers through society*. Cambridge: Harvard.

Latour, B. (1993). *We have never been modern*. Cambridge: Harvard.

Latour, B. (1999). *Pandora's hope: Essays on the reality of science studies*. Cambridge: Harvard.

Loving, C. (1997). From the summit of truth to its slippery slopes: science education's journey through postivist-postmodern territory. In *American Educational Research Journal* (34:3): 421–452.

Maturana, H., & Varela, F. (1987). *The tree of knowledge: The biological roots of human understanding*. Boston: Shambhala.

Messer–Davidow, E. (1995). Manufacturing the attack on liberalized higher education. In Christopher Newfield and Ronald Strickland, eds., *After political correctness: The humanities and society in the 1990s*, 38–78. Boulder: Westview.

Percival, I. (1991). Chaos: A science for the real world. In Nina Hall, ed., *Exploring chaos: A guide to the new science of disorder*, 11–21. New York: Norton.

Plotnitsky, A. (1994). *Complementarity: Anti-epistemology after Bohr and Derrida*. Durham: Duke.

Plotnitsky, A. (1993). *Reconfigurations: Critical theory and general economy*. Gainesville: University of Florida.

Prigogine, I. (1996). *The end of certainty: Time, chaos, and the new laws of nature*. New York: Free Press.

Ross, A. (1996). Cultural studies and the challenge of science. In Cary Nelson and Dilip Parameshwar Gaonkar, eds., *Disciplinarity and dissent in cultural studies*, 171–184. New York: Routledge.

Ross, D. (1991). *The origins of American social science*. Cambridge: Cambridge University.

Scholes, R. (1998). *The rise and fall of English: Reconstructing English as a discipline*. New Haven: Yale.

Sokal, A. (1996). A physicist experiments with cultural studies. In *Lingua Franca*, 62–64.

Sokal, A., & Bricmont, J. (1998). *Fashionable nonsense: Postmodern intellectuals' abuse of science*. New York: Picador.

Stewart, I., & Cohen, J. (1997). *Figments of reality: The evolution of the curious mind*. Cambridge: Cambridge University.

Traweek, S. (1996). Unity, dyads, triads, quads, and complexity: Cultural choreographies of science.In Andrew Ross, ed. *Science Wars*, 139–150. Durham: Duke.

Usher, R., & Edwards, R. (1994). *Postmodernism and education*. New York: Routledge.

Part One
Science as Curriculum Theory
Marla Morris

William E. Doll, Jr.'s essay entitled "The Object(s) of Culture: Bruno Latour and the Relationship Between Science and Culture" explores the difficult terrain of Bruno Latour. Accordingly, Doll suggests that Latour is concerned that bifurcations, and what Latour terms "purifications," of the disciplines, wedges driven between the scientific and the social, are misguided. Rather, Latour calls for the recognition of "hybrids." Doll claims that Latour calls for the mixing of premodern and modern sensibilities. Here, hybrids take on their own life. Interestingly, Doll makes a connection between Latour's work on technology and Heidegger's "The Question Concerning Technology." Doll suggests that Latour's work on technology is an "extension" of Heidegger's thinking on the subject. The complexities of culture and technology, the ways in which technology, culture, and politics are interwoven, are not, Doll suggests, easy to determine. Supplementing William Doll's chapter, Franc Feng and Stephen Petrina suggest in their postscript, "Cosmopolitically Speaking," that Latour, drawing on Isabella Stengers, calls for the hybridization of the epistemological, ethical, ontological, political, and theological. Further, a cosmopolitic calls for "coexistence," "participations," and "collectives" that allow humans and nonhumans to live together.

Eva Krugly-Smolska, in her essay called "Bachelard as Constructivist," suggests that Bachelard calls for "contextualized" knowledge. Decontextualized knowledge, or formalism, simply will not do. Bachelard, says Krugly-Smolska, raises some interesting questions around how we come to know things scientifically. "Epistemological breaks" often come by way of "hindrances" and "obstacles." However, there is no "unitary epistemology" of scientific thought. Further, according to Krugly-Smolska, Bachelard suggests that developments in scientific thinking do not break through to the new, but rather what transpires is an "envelopment of old thought by [the] new."

David Blades, in his essay entitled "The Simulacra of Science Education," explores the work of Jean Baudrillard. Blades suggests that the key issue for Baudrillard turns on the notion of reality. As against correspondence theories, whereby a representation of something captures the thing and names it as reality; Baudrillard suggests that re-presentations cannot do this. Signifiers only signify more signifiers. Thus the quest to find reality is nothing more than a "game." Blades' reading of Baudrillard suggests that signifiers not only "make reality disappear" but also "mask that disappearance." This is what the notion of simulacra means. This would be the "perfect crime." Signifiers do not represent the signified. They re-present only themselves. The "hyperreal," then, serves to put "distances" between what we see and what we get. Blades asks interesting questions around what it is that science education is doing in light of Baudrillard's claims.

Marla Morris' essay, entitled "Serres Bugs the Curriculum," explores connections between Serres' philosophy, information theory, and premodern philosophy. Morris' thesis is that Serres engages in an "onto-theology." Being (ontology) is not substance, but noise. Noise is a kind of interference (like the interference in information theory, i.e., the interference in a phone line) that signifies primordial chaos; noise is a kind of theology. Noise is connected to something divine. The divine is not silent and not transcendent; the divine is, rather, noisy and immanent. Noise (Being) is further not just an ontological principle that sits on high and watches from above, it is embedded and embodied in beings (entities) such as bugs (parasites) and ghosts (the Paraclete). How do these beings know? What kind of epistemology is at hand? Serres tells us that beings know in a relational, fragile epistemology. Parasites can get their feelings hurt. No matter, when they walk into the room, Serres tells us, city rats and country rats are changed, disturbed, whether they like it or not. Serres offers a "mystical poesis" says Morris.

Chapter One
The Object(s) of Culture: Bruno Latour and the Relationship between Science and Culture
William E. Doll, Jr., Franc Feng, and Stephen Petrina

Bruno Latour, the French contemporary sociologist, anthropologist, philosopher, commentator on current cultural trends, begins one of his recent books, *We Have Never Been Modern* (1993), with an analysis of the newspaper stories he and his colleagues find daily. These stories—about global warming, AIDS vaccines, frozen embryos, the Pope and contraception pills—he calls "hybrid articles" (p. 2). They are hybrids because, neither pure science nor pure politics, they muddle and mix, in an ever-weaving network of social and technical relations, the cultural and the technological, the social and the scientific. To know how to handle the issues raised by global warming, AIDS vaccines, frozen embryos, and contraception pills is virtually impossible, for we have no past history of dealing with such hybrids. The purity of our traditional disciplines has not prepared us to deal with these hybridized networks, themselves a product of our unmodern society where "science, politics, economy, law, religion, technology, [and] fiction" (p.2) mix and match themselves in increasingly complex displays. But dazzling as are these hybridized networks—since they take on a life and persona of their own, Latour calls them "quasi-objects" (p.51 ff.)—they really have no history, no upbringing, no parentage, either scientifically or socially. Thus, they are dazzling and dangerous, *enfants terribles*.

While reading Latour on these "objects," I decided to pick up my own newspaper: Behold, I saw headlines describing the latest version of the U.S. Congress' current attempt to establish a "National Defense Shield" (*Times-Picayune*, March 25, 1999). This shield, euphemistically called "Star Wars," was first suggested and supported by Ronald Reagan when he was president. The article, on the front page, intermixes technology, politics, economics, and patriotism by having both Louisiana senators make comments on its prospects, showing graphic drawings of how American radar would 1) detect enemy missiles, 2) distinguish between warhead missiles and decoys, and 3) launch the

U.S.'s own interceptor missiles to 4) destroy the incoming warheaded missiles. Part of the comments read that this newer, updated version, quietly under development for the past ten years, will be able " to defend against small-scale missile attacks," launched by a "rogue nation." The very last sentence, under the four-stage graphics, and in small print reads: "The concept is similar to star wars; the program's feasibility remains unclear" (p.A6). The creature lives but who are its parents? Who gave it birth? Scientists, technicians, politicians, patriots? And who will provide the (financial) sustenance needed for its nurture?

In his latest (translated) book, *ARAMIS or the Love of Technology* (1996), Latour explores the same questions posed by Star Wars. ARAMIS, obviously a play on one of Alexandre Dumas' "Musketeers" is an acronym for *Agencement en Rames Automatisées de Modules Indèpendants dans les Stations* or Arrangement in Automated Trains of Independent Modules in Stations—more colloquially, small, mass transit cars, running in tandem but without being hitched mechanically. The idea is that, as independent vehicles, connected via "nonmaterial coupling" (magnets or electricity), the cars rather than the people in the cars would do the switching as the car-trains moved about the city. This idea caught people's—engineers, city planners, politicians—imagination and so, conceived in 1969, ARAMIS was born (received funding) in 1972, led a wandering youth without a definite family, was confirmed (received new bureaucratic blessing) in 1984, and quietly and silently passed from the life in 1987, still just a youth. Latour's quasi-novel is about "Who Killed ARAMIS?" In a sense no one killed him (Yes, ARAMIS was a male quasi-object) for he was at best, a dream and what protoplasm did exist became diffused among a number of other agencies and projects. As the sociologist and his young engineering assistant do a postmortem (a postmodern postmortem?) to discover the cause of "death," they find not real body, just bits and pieces of old records, letters, hopes. One government minister says "It was a seductive idea, ARAMIS—really quite ingenious" (p.9). Aramis, speaking beyond the grave on his own behalf, says later in the book, "You loved me as an idea. You loved me as long as I was vague," unreal (p.294). Did ARAMIS' death actually lie in his birth? Does our love of technology always need to lie in our dreams (of what it possibly can do)? Does the birth of that technology always disappoint? Think of the combustion engine! Nuclear power! "Smart bombs"! Yet, yet, yet...the hope remains!

As the book ends (pp.300–301), the sociologist (Latour) is disillusioned, but the young engineer (technology's new man) decides to

journey off to find a "purely technological project, a doable project." He picks "smart cars," ones that run individually and automatically over technologically developed highways. On the way back from a "smart car" conference in California (where else!) he notices an article in the *San Diego Union*: FAMILY–SIZED MASS TRANSIT CARS TO BE STUDIED AS ANSWER TO CITY CONGESTION.
He says to himself:

> Damn...if they'd just waited...Aramis would have been on the right path...A billion dollars...It's all becoming profitable again. I should have stuck with guided transportation. (p.301)

Latour's intent is not to denigrate technology, far from it; technology not only is part and parcel of our current life, underpinning that life, as it were, it also is an expression of the procreative urge with us. Rather Latour seeks "to show technicians...that by becoming good sociologists and humanists they can become better engineers"; and to show sociologists that "they can welcome crowds of nonhumans with open arms." (p.viii). Latour wants to bridge or fuse two "clearly separated universes." Unless we understand how research melds with politics, unless we understand how both are fueled by the human passion to create, unless we understand the cultural origins of technology, and how technology represents the hopes and consequently the failures of our society (Martin Heidegger's concept of *Gestell*, 1977/1954), we will destroy ourselves as we promiscuously conceive, unconsciously birth, uncritically nurture, and blindly follow the quasi-monsters we have created.

ARAMIS (1996) is a playfully serious book; all the facts in the novel are true but they are placed in a fictitious frame; thus "the hybrid genre I have devised for a hybrid task is what I call *scientifiction*" (p. ix). *We Have Never Been Modern* (1993) is playful in its own right, but it is also deadly serious, as ARAMIS is only serious; *We Have Never Been Modern* is an academic book and represents Latour's (currently translated) best attempt at wrestling with the issues of science and culture, particularly in our "post" age. There are a number of ways to view Latour's arguments in this book, and one of them is certainly through the concept of our "post" age, an issue Latour obliquely refers to in his title. While I will comment on this strand later, I'd like to begin with Heidegger's *Gestell*. In some ways it is possible to see Latour's own views on technology and society not only being influenced by but being an extension of Heidegger's provocative and seminal, 1950s essay, "The Question Concerning Technology" (1977/1954).

In this essay, Heidegger posits that there lies in technology a supreme danger, indeed *the supreme danger* (p.26). But he also says, throughout the essay, that "The essence of technology is by no means anything technological" (p.4), that "The essence of modern technology…is itself nothing technological" (p.20), and that "What is dangerous is not technology" (p.28). That is, the supreme danger which frightens Heidegger so much lies not in technology itself, nor really even in our relationship with technology, but in *the essence of technology*—its *Gestell*. *Gestell*, Heidegger says, means in its ordinary usage, a frame, "some kind of apparatus, e.g., a bookrack" (p.20). But in speaking of *das Ge–stell*—"the name for the essence of modern technology" (pp.21 and 20), Heidegger has turned the simple sense of frame into a complex sense of "enframing." In enframing, we come in contact with "the sense of destining and danger" (p. 28). What Heidegger is after here is that modern technology—and technology is a way to define "modern man"—as a vehicle, enframes, entraps, seduces us into a particular *way of thinking*. This way of thinking enframes us in a mode whereby

> man…exalts himself to the posture of lord of the earth. In this way the impression comes to prevail that everything man encounters exists only in so far as it is his construct. (p. 27)

Since technology is so obviously a human construct, indeed a human male construct, its very success—nuclear power, "ethical" drugs, advanced weaponry—seduces us ("challenges us forth" really) to think *only* in technological terms. In so doing we become not only enframed but also entrapped—as is the young engineer in ARAMIS—and thus conceal from ourselves our human sense of *being*. As Heidegger puts it:

> The rule of Enframing [*Gestell*] threatens man with the possibility that it could be denied to him to enter into a more original revealing and hence to experience the call to a more primal truth. (p. 28)

Heidegger's fear, one shared by Latour, who mentions Heidegger a number of times and *Gestell* at least once (1993, p.124), is that in relying on technology and its frame of thought we lose sight of our basic humanness—we destroy villages and those in them in order to "save" such for democracy. Within this mechanistically ordered, "cause-effect coherence [says Heidegger] even God can…lose all that is exalted and holy" (p.26).[1]

The Object(s) of Culture

Latour explores, with depth, this relationship between democratic humanness and technological "rationality" in *We Were Never Modern* (1993). He argues that these two come from differing (and often unrecognized, concealed) strands within the "Enlightenment"—a scientific strand and a humanist strand. The word "modern," referring to "the historical period that is ending," (p.11) "designates two entirely different practices" (p.10)—one set of practices being pure and objective, the other being mediated and subjective. The first, the scientific strand, is based on the idea of a "transcendent Nature" (p.41), one that can be studied (and indeed conquered) apart from human peccadilloes; the other, the humanist strand, is based on the idea of a mediated nature, culture in its various ways and forms. Not only have the moderns, "the mods," carefully separated these two—Nature and Society—they have polarized this separation and "credit only the former" with whatever success they might consider to have achieved over the past centuries (p.41).

For Latour this dichotomization, so characteristic of modernism but so little understood by the modernists—hence the title of never being modern—has produced a "crisis": the proliferation of hybrids. These hybrids, which have increased dramatically in the past decades, result from interbreeding of the scientific and the social: "frozen embryos, expert systems, digital machines, sensor-equipped robots, super vegetables, data banks, gene synthesizers" (p.49). Like ARAMIS they do have a life of their own, but they are not purely the children of science nor of society; the separation between these two is so great that science and society do not realize they have united to produce children. These children are thus left on their own, to wander the world without family, without parents. They are in a word, "monsters." We the people, who have the ultimate responsibility for dealing with these hybrids, have no training in such, believing in the myth of separateness, of the "absolute dichotomy" (p.40) between science/society, technology/culture.

We cannot be responsible—to the natural world we inhabit, to ourselves as humans who inhabit—as long as we continue to see these two worlds and their habits of thought and patterns of action as absolutely separated. We have written, at least tacitly, a constitution of rights and responsibilities based on this separation. We need a new constitution, one which sees that the work of purification is and must be intertwined with the work of mediation, that the technological and scientific are influenced by and also influence the cultural and social. This understanding is not one the young engineer in ARAMIS was ever able to achieve; the story ends with his still believing in the "purity" of science and its handmaiden technology, that with sufficient resources (more bombings in Kosovo) a better world will be created for all.[2]

In his call for a new constitution—an official, recognized, mutually agreed–upon way of looking at the world and its proliferation of hybridized networks—Latour is by no means denigrating science and its technological accomplishments. Far, far from it. He chastises both Habermas (p.60) and postmodernism (p.46) for their rejection of empirical research. As he says, "Instead of moving on to empirical studies of the networks that give meaning[3]...postmodernism rejects all empirical work as illusory and deceptively scientistic" (p.46). Postmodernism is for Latour more a "symptom" "of something gone awry" than it is "a fresh solution" (p.46). Postmodernism (at least that of the nonchaotic variety),[4] Latour believes, remains "suspended between belief and doubt, waiting for the end of the millennium" (p.9). He quite firmly dismisses this movement. Instead he wants us to explore a new land, the land of hybridized networks, the land "of nonmodern worlds. [Here lies] the Middle Kingdom [one where science and society not only affect one another but are understood to do so] as vast as China and as little known"(p.48).

This Middle Kingdom that Latour wishes to resurrect is not a totally new kingdom, it is one which has roots in the premodern, that world which existed before modernism bifurcated all (purifying, objectifying, and giving precedence to the scientific). This land needs a new constitution, one which recognizes not just people but *things*, the objects of science and the quasi-objects bred by the union of science and society. "We want the meticulous sorting [and representing] of quasi-objects to become possible" (p.142); "The imbroglios and networks that had no place...are the ones that have to be represented, it is around them that the Parliament of Things gathers henceforth" (p.144).

In this Parliament of Things, governed by the unmodern constitution, in this land of the Middle Kingdom, Latour is calling for two "newnesses." One is to put these objects and quasi-objects on display for full viewing and debate—the ozone hole itself should have representation, as should the chemical industry, and as should those working in that industry, and as should the residents in the states immediately affected by the meteorology of the polar regions (p.144). That which is now discussed solo voice, placed under the table, and covered with specialized language, should be placed on the table, discussed openly, honestly, and directly in terms the citizenry can understand. The notice that the Star Wars defense missile system has not been perfected should appear upfront, not as a small and final squib

The Object(s) of Culture 31

placed under the graphic. What "Is" should be evident, not camouflaged or hidden.

The second "newness" is that in re-conceiving the concept of change (the charge as Latour sees it, p.145), we should not be caught in modernism's bifurcation mode; thus we should not remove ourselves from all that is premodern or modern, or even postmodern. All of these have approaches to issues that are worth keeping: From premodern thought we should keep a sense of unity and integration, their certainty that transcendence abounds. From the moderns we should keep their proliferation of hybrids, their ever-increasing scale of action, their creation of stabilized objects, their sense of freedom. From the postmoderns we should keep their sense of incredulity, their pluralism, their reflexivity (pp.133–135). From the sciences we should keep "their daring, their experimentation, their uncertainty, their warmth" (p.142). But we will no longer believe in their exclusiveness, "their objectivity, their truth, their coldness, their extraterritoriality—qualities they have never [really] had" (p.142). This land of the Middle Kingdom—between the poles of objectivity and subjectivity, absolutism and relativism, science and society—will be a brave and true new land.

What I believe Latour is after here is not just a unification of that which modernism has rent asunder (in theory, not in practice). Latour underestimates himself when he says: "I have simply reestablished symmetry between the two branches of government, that of things—called science and technology—and that of human beings" (p.138). Latour's reestablishment of symmetry is more than a simple bringing together; he is asking for an "amalgam" of the premodern and modern, a blending together or folding into one another. In this amalgamated process, the uniqueness of each retains its own flavor while at the same time, through interconnections, being an integral part of a larger network. Without the uniqueness of the science we have we would not have the society we have. Each is the other to/of the other; here is a new way to visualize self. We are each of us, in our own selves, the other to our other. Our selfness depends on the quality of our otherness.

In this process of building a "nature-culture" network (p.7), Latour intends to keep and maintain the autonomy of both science and society, while at the same time blending them into a new whole, a whole where all have voices.

Cosmopolitically Speaking: A Postscript

How can we "become once more what we have never ceased to be"? How can we become amodern if we never ceased to be amodern? This is the riddle which Latour (1993) resolves. Is the answer to this riddle to be found in two modern practices—purification and mediation? Evidently so. Latour cloaks himself as an amodern anthropologist, demonstrates the collusion of these separate practices, and unravels the moderns' relations between the divine, natural, and social. He resolves the riddle through a simple, but thorough, revision of history (pp.40–43). Modernity is a notion we used to understand what was happening for the past three or four hundred years and to influence our fate. And so, like modernity's child, progressivism, into the dustbin goes the parent. Modernity as a reified way of being is to be distinguished from modernism as a mode of practice. Then what is or was this thing called modernity? Modernity, as Latour sees it, amounts to a constitution of guarantees for practicing the likes of civil and natural law, politics, religion, science, and technology. Modernity's power comes through certain checks and balances (relations) among the guarantees, which are written as follows: "[Moderns] have not made Nature; they make Society; they make Nature; they have not made society; they have not made either, God has made everything; God has made nothing, they have made everything" (p.34). For example, within this matrix of relations, we can practice science and remain deeply spiritual: we can explain natural or social events without ever appealing to divine intervention. Or we can accept that nature is real while believing that social power is constructed. In one breath, the constitution guarantees this distinct separation or *purification* of God, Nature, and Society (or the divine, natural, and social), the relations among them depending on whether the three are transcendent or immanent. In the same breath, the constitution renders the *translation* or *mediation* of these into hybrids (mixes of humans, nature, objects, societies, and God/s) "invisible, unthinkable, unrepresentable" (p.34). If modernity meant that we purify God, Nature, and Society as we eliminate mixtures of these three, then we were never modern.

Like the antimods and postmods, Latour is suspicious of modernity and its pretensions, but he does not share their pessimism, nihilism, or solipsism. Here is Latour at his eloquent best. Locating himself in position outside the three modern critiques, yet in the same breath engaged in a combined variant of these very three criticisms (although not by name), Latour cleverly indicts modernity from an almost unassailable contradictory position. He is an anthropologist from the "not-so-distant" past, which he claims within the same breath—has never passed. In so doing, Latour trumps the antimods and postmods in expressing his own suspicion of modernity. Moreover, Latour makes his

The Object(s) of Culture 33

claims, through a shared antimony of the notion of progress, and irreversible time—but without "negations" of the two groups that he both identifies and classifies—neither the negativism of the antimods, nor the paralysis of the postmods. More strategically, Latour couches his claims within a convincing thesis. Through joint but unacknowledged practices of purification and mediation we have proliferated in creating hybrids of the three without representation. The unrepresented masses (of hybrids) saturate the modern constitution. To deal with this fact that we have never been modern, Latour, with all the masses he can mobilize, is rewriting the constitution.

"No Mediation Without Representation"
The first draft of the amodern constitution attends to the work of mediation, representation of hybrids, and the protection of rights. The constitutional goal is to protect the rights of human, nonhumans, and hybrids of the two. The first guarantee protects the work of mediation. Says Latour, "every concept, every institution, every practice that interferes with the continuous deployment of collectives [i.e., natures-societies] and their experimentation with hybrids will be deemed dangerous, harmful, and—we might as well say it—immoral" (p.139). The second guarantee protects the transcendence of nature and of society: "All concepts, all institutions, all practices that interfere with the progressive objectivization of nature,...and simultaneously the subjectivization of society—freedom of maneuver—will be deemed harmful, dangerous and, quite simply, immoral" (p.139). This second guarantee places a check on the first by recognizing differences within hybrids between natures that we have not made and societies that we are free to change. The third guarantee attends to the work of mediation by protecting the free assembly of hybrids without having to represent things as either cultural or natural, human or non-human, local or global, old or new, etceteras. And the fourth guarantee is to "replace the clandestine proliferation of hybrids by their regulated and commonly-agreed upon production" (p.142). This last guarantee, says Latour, acknowledges the dangers of a proliferation of hybrids consisting of objects and subjects which have no representation. Latour envisions an expanded democracy where not only everyone, but also everything, is represented. The four guarantees protect our rights to mediate, delegate, participate, and regulate, more or less in that order; in short, they protect our rights to legislate. These guarantees also reconstitute the divine or supernatural; transcendences abound as our world was never any more disenchanted than we were modern. When transcendences abound, are

not relations between what we represent as natural and as social, mediated? Are not fully naturalized or socialized (or politicized) worlds improbable? What is guaranteed is an amodern cosmopolitics of checks and balances among hybrids of what we choose to represent as divine, natural, and social. Nietzsche's pronouncement of the 'death of God' and the tales of disenchantment that followed were modern self-fulfilling prophesies. Our cosmos is enchanted.

With constitutional guarantees extended to artifacts and natural things, nature is not unified, culture not limited to symbols, and God is not "up there," impotent yet at the same time sovereign. The world is not divided across so many cultures (which can be modified at will) and one nature (which is durable). An amodern constitution prompts us to speak in terms of monoculturalism and multinaturalism. Multinaturalism suggests "nature" is not a unified and universal whole and monoculturalism in turn opens up a dialogue on what we have in common. Why, Latour asks, do we think of nature in the singular and of culture in the plural? "What do we have in common?" "In which common world are we ready to coexist? Who will define the common world?"

Are We the People Cosmopolitically Correct?
What will an anthropology of the a/modern east look like to an eastern amodern ethnographer? Will the title of her ethnography be *We Have Never Been Colonial?* What will an anthropology of a/modern west look like to an eastern amodern ethnographer? Will this have to be titled *You Have Never Been Modern?* Latour does not so much avoid identity politics inasmuch as he reconstitutes identities and representations through cosmopolitics. Do not confuse Latour's "We" in *We Have Never Been Modern* with the "royal we" nor even the "self-modern we." Rather, Latour speaks *on* the "collective we"—on humans and nonhumans (Latour, 1993). While social histories cannot be unified, a history which immerses us all as a "collective-in-the-world" opens a different conversation. Cosmopolitics attends to the mediation of conjoined fates, and interests in a common home that "is not of our making."

In her work on the collective we, says Latour (1997), Isabelle Stengers has been able to so thoroughly mix epistemological, ethical, ontological, political, and theological components that it can only be what she calls it: cosmopolitics. This thorough mixing—this mediation of passes—this legislation of cosmos—opens different "regimes of encounter" as it takes nothing for granted. The problem is no longer whether something is PC (Politically Correct) in identity politics, but whether it is CC (Cosmopolitically Correct) in cosmopolitics. For example, we can exercise our rights in the new constitution in a way that we have put nothing at risk; we realize only a reconstitution of the status quo. Rather, say Latour and Stengers, we ought to use our guarantees to constitute hybrids where we and the status quo are at risk; this would be cosmopolitically correct.

To be sure, a delegation and representation of things are a risky cosmopolitics. From a standpoint of identity politics, these are nearly unassailable. Yet, it is precisely identity conceptions of exclusion and inclusion that Latour and Stengers have subverted. In place of these conceptions are participations, represented in varied forms. Representations are governed by participations among humans and nonhumans within hybrids or collectives, or by what anthropologists have called the *law of participation* (Levy-Bruhl, 1926/1979, pp.69–104). Everything participates, not only through static connections, but through circulation of influences, powers, qualities, virtues, etceteras. We can be felt outside and at the same time remain where we are. Indeed, this is not to be understood as a share or a fraction of the whole of properties, nor as an organ in an organism. We need not affirm oppositions between one and the many—the individual is in the collective and the collective is in the individual. The verb "to be" has no ordinary correspondence to identity, and encompasses collective representation. Any good legislator dare not ignore the law of participation: if everyone and everything participates, then all ought to be represented.

How do we go about representing things that participate? Latour envisions two chambers in his Parliament of Things (Latour, 1998, 1999). 'How many are we?' is asked in one chamber. The first two requirements, perplexity and consultation, are stated as follows: "Thou shall not simplify the number of entities to be taken into account (Perplexity). And Thou shall check if the number of voices in the discussion about new entities has not been arbitrarily short cut (Consultation)." 'Can we all live together?' is asked in a second chamber, attending to representation as definition. The second two requirements, hierarchization and certainty, are stated: "Thou shall establish the hierarchy of new entities with those already in place (Hierarchization). And once the entities are institutionalized through legitimate presence in the collective, thou shall not discuss any longer (Certainty)." Through perplexity and consultation, we take into account who (human and nonhuman) should vote and have voice. Through hierarchization and certainty, we make certain we have phenomena ordered by importance. Cosmopolitics becomes a question of coexistence.

"United We Stand in Nature, Divided We Become Understandable"
Unlike the metaphysicians who prescribe the rules of life for individual enlightenment, the cosmopolitician Latour is legislating for collective constitution. Through Latour's legislation, he places the cosmos at risk, and notable discomfort is felt by all of us. So what are we to do? Latour is legislating—he may be a cosmopolitician but he is certainly *not* a prophet. He does not offer a path to follow, but at the same time, he is not inclined to say that we make our own individual paths. As we discover collective potentiality, we work out our own potentiality. Latour only reminds us: We work this out as a collective of natures-societies. Left originally divided, we (humans and nonhumans) were understandable only in cultures, but united we stand in natures.

Endnotes
1. Heidegger does not end his essay on a purely negative note—one wherein "Enframing reigns," thus "concealing" all *Being* except the techno-rational, and leaving us helpless against its "destining" power. Rather, just after talking in these terms he quotes the following line from the German poet Friedrich Hölderlin: "But where danger is, grows the saving power also." (p.28) This saving power is that of "man's" artistic being, his sense of *poesis*. Through "thinking"—thinking which questions reflectively and deeply—are we able to "keep watch over the unconcealment" (p.32) which entraps us and makes us unwitting slaves of the need to control. In seeing "technology as an instrument [by which we can control nature for society's "best" ends] we remain held fast in the will to master it" (p.32). This is the "supreme danger," the one concealed from us by the "essence" of technology: the more we think in control terms, the more we unwittingly become slaves of the need to control, and the more we see Being only in terms of control.

Poesis represents that aspect of Being wherein the artistic, aesthetic, human, beautiful, spirit(ful), and true blend. The ideal is a conjunction of *techné* and *poesis*, and our power to bring forth this ideal occurs just at the darkest moment, when *Gestell* is the strongest. But it will occur only if we are ever vigilant not to get caught up in the active frenzy of technology and in the seduction of its productivity—as did the young engineer in ARAMIS. Instead, we need to recognize that spiritful and contemplative "questioning is the piety of thought" (p.35). Grand as this idea is, will it ever happen? Heidegger says, "Whether art may be granted this highest possibility of its essence in the midst of the extreme danger, no one can tell" (p.35).

The Object(s) of Culture 37

Richard Bernstein (1993, Ch.4), while praising this ideal, theoretically, also sees its practical dangers and how these dangers played themselves out in Heidegger's own life, especially in his exalting of the poetic (*poesis*) over the practically active (*phronésis*) and thus succumbing to the transcendent lure of Nazism.

2. It is interesting to note that the conflict in Kosovo is just the sort of quasi-object (network of hybrids) that Latour is bringing to our attention—it is not military in the usual sense for it has been predicated on advanced technology winning all. That (unexpectedly) failing, personal negotiations were integrated with the bombing. No ground troops, the heart of all previous military operations, have been used. This is a new "war," one quite disconcerting to the generals in charge and one which will require a rewriting of textbooks on military strategy. The battle plan was predicated on the complete success of high technology (conventional helicopters, although potent, were never used), and that failing, negotiations were introduced. The technological and the social have influenced each other with the strategy shifting from technology alone to technology with negotiations (and negotiating with an indicted war criminal has been anathema to many military personnel). Working this hybrid network has been disconcerting to many, especially to those wishing to use technology in greater degrees (broader, less selective target bombing and more of the bombing) and those who wished to use negotiations exclusively or at least with bombing only as a threat.

As one speculates about the use of our most high-powered military technology, atomic bombs, one wonders whether in World War II we would have used such against Europeans as opposed to our use of such against "Japs." Are Serbians European? I believe Latour is correct, the hybridization of culture and technology is intricate in ways not easily seen. Does not war and the killing of the "other" always require us to see the "other" not just as other but as a foreign, barbarian "other"? And do we not need to use social manipulation, strongly influenced by language, to achieve this purpose? Where does truth reside in this maze?

3. In the phrase "empirical studies of the networks that give meaning," I am quickly reminded of the work of Gregory Bateson. He, too, saw mind and nature as a necessary unity, as a network of relations. See *Mind and Nature* (1979), *Angels Fear* (1987), and Morris Berman's work on Bateson, *The Reenchantment of the World* (1981, Chs. 7, 8,9).

4. While Latour sees postmodernism as a confused movement—a "hyperreality [where} nothing has value; everything is a reflection, a simulacrum, a floating sign" (p.131), he does recognize that postmodernism is struggling with the contradictions of modernism—"the

postmoderns have sensed the crisis of the moderns and attempted to overcome it" (p.134)—and sees that in their "pronounced taste for reflexivity" there are attributes worth keeping. This is especially true of the "chaotic" [my word] postmodernism of Michel Serres, whom Latour praises extensively (pp.52, 84). I, too, believe there is much to admire in this quite overlooked branch of postmodernism (Doll, 1993, 2000).

References

Bateson, G. (1979). *Mind and nature: A necessary unity*. New York: Dutton.

Bateson, G., & Bateson, C. (1987). *Angels fear: Toward an epistemology of the sacred*. New York: Macmillan.

Berman, M. (1981). *The reenchantment of the world*. Ithaca: Cornell.

Bernstein, R. (1993). *The new constellation: The ethical– political horizons of modernity/postmodernity*. Cambridge, MA: MIT.

Doll, W. E. Jr. (1993, 2000). *A post-modern perspective on curriculum*. New York: Teachers College.

Heidegger, M. (1977). The question concerning technology. In *The question concerning technology and other essays* (William Lovitt, trans.). New York: Garland (pp.3–35). (Essay, original German publication, 1954).

Latour, B. (1993). *We have never been modern* (Catherine Porter, trans.). Cambridge, MA: Harvard.

Latour, B. (1996). *ARAMIS or the love of technology* (Catherine Porter, trans.). Cambridge, MA: Harvard.

Latour, B. (1997). Foreword: Stengers's shibboleth. In Isabelle Stengers, *Power and invention: Theory out of bounds* (pp.vii–xix). Minneapolis: University of Minnesota.

Latour, B. (1998). From multinaturalism to politics. Paper presented at the University of British Columbia, 4 November 1998.

Latour, B. (1999). *Pandora's hope: Essays on the reality of science studies*. Cambridge, MA: Harvard.

Levy-Bruhl, L. (1926/1979). *How natives think*. New York: Arno.

Times-Picayune (March 25th, 1999). Missile-defense bill updated 'Star Wars.'(Section A1 and A6), New Orleans, Louisiana.

Postscript

For those interested in how Latour has influenced the field of education, let us suggest the following:

Fountain, R. (1998). Sociologics: An analytical tool for examining socioscientific discourse. *Research in Science Education*, 28(1), 119–132.

Gaskell, J., & Hepburn, G. (1998). The course as a token: A construction of/by networks. *Research in Science Education*, 28(1), 65–76.

Hepburn, G. (1996). Working the net: Initiating a new science and technology course. (Ph.D. dissertation, University of British Columbia).

Nespor, J. (1994). *Knowledge in motion: Space, time and curriculum in undergraduate physics and management.* New York: Falmer.

Rafea, A. (1999). Power, curriculum making and actor–network theory (ANT): The case of physics, technology and society curriculum (PTS) in Bahrain. (Ph.D. dissertation, University of British Columbia).

Chapter Two
Bachelard as Constructivist
Eva Krugly-Smolska

Gaston Bachelard came to the philosophy of science relatively late in his career. He was working as a telegraphiste with an intention of becoming an engineer, when World War I broke out. On his return from the front he started teaching physics and chemistry at his old high school. It was at the age of thirty-six that he started studying philosophy while continuing to teach. He received his doctorate at the Sorbonne in 1927 at the age of forty-three. A year later with the publication of his two theses, his career in philosophy had finally begun. He would return to the Sorbonne in 1940 as a professor of the history and philosophy of science and remain there until his retirement in 1955. During this period the science-poetry polarity of his work was very much in evidence.[1]

Bachelard was a prolific writer with a legacy of over ninety publications, which includes twenty-three books, among them twelve on the philosophy of science. While he is known to every high school student in France, he is little known to the English speaking public, mostly because of the limited number of translations available, which are mostly of his works on poetry.[2] Recent interest appears to be as a result of his influence on other French philosophers such as Foucault, Derrida, Althusser, and others for whom he is the inventor of the concept of "epistemological break." What is new about twentieth century science for Bachelard is that scientific progress always reveals a break between ordinary knowledge and scientific knowledge. No longer does it have foremost an empirical basis; it has also broken with deductive logic. It is no longer Cartesian (McAllester Jones, 1991).

Given the extent of his work, it is impossible to do justice to Bachelard's work in the limited confines of this one chapter. Instead, a very limited focus will be taken, that of Bachelard's notions about science and pedagogy. Bachelard the science teacher is very much in evidence in Bachelard the philosopher of science, as the title of one of his books, *La Formation de l'esprit scientifique* (1996 [1938]), indicates. In French, *formation* is often used where we would use the word education. However, the term can also mean structure or structuring. Such translation problems arise also with respect to *esprit scientifique*, which can be translated either as 'scientific mind' or 'scientific spirit' (and has been). In both cases both meanings are appropriate but at different times as will become significant in examining his epistemology and his approach to the education of the scientific mind. The uppermost question will be whether Bachelard can be considered a constructivist epistemologically and/or pedagogically. The three books which will be the focus of this discussion will be *Le Nouvel esprit scientifique* (1934), *La Formation de l'esprit scientifique* (1996 [1938]), and *La Philosophie du non* (1981 [1940]), henceforth referred to as NES, FES, and PN respectively. The last will be referred to only briefly in order to introduce a concept, the epistemological profile, which is central to the issues of constructivism and pedagogy, with which this chapter is concerned.

In NES, Bachelard tells us that "science is a product of the human mind, produced in accordance with the laws of our thought and adapted to the exterior world. It offers, therefore, two aspects, one subjective, the other objective, both equally necessary" (p.2).[3] As a result he considers the dichotomy of realism and rationalism as false, arguing that both are always present although one or the other may be dominant at any point in time. "If scientific activity is experimental then reasoning will be required; if it is rational then experiment will be required" (p.3) He further argues that neither is absolute and that one should not start with a particular philosophical position to judge scientific thought. On the contrary, "science in effect creates philosophy" (p.3). Nevertheless, there is an epistemological vector which moves, according to Bachelard, from the rational to the real and not the inverse, as philosophers from Aristotle to Bacon would have it.

It should be remembered that for Bachelard contemporary science is quantum physics. The dialectic that he sets up between realism and rationality is, therefore, perhaps not surprising. Indeed, he tells us that one can't help but notice that dialectical tendencies in philosophy appear at more or less the same time as they do in science (p.20). Because his focus is mostly on the physical sciences, it is no surprise that mathematics is brought into the equation. "Mathematical realism will sooner or later give body to thought, giving it psychological permanence, finally splitting mental activity, making appear, here as everywhere, the duality of the subjective and the objective" (p.5).[4] This is a second-order realism, a technical realism, one that arises from reason that has been realized, reason tested by experiment. "A scientific experiment is reason that has been confirmed. This new philosophical aspect of science is paving the way for the return of the normative to its experiments; since the need for an experiment is grasped by theory before it is discovered by observation...reasoning by construction is now making its appearance in both mathematical and experimental physics" (pp.5–6).[5] "From now on, hypotheses are syntheses" and "mathematics opens new ways for experimentation." Indeed, "it is a mathematical instrument that creates contemporary physical science in the way that the microscope creates biology; no new knowledge without mastery of this new mathematical instrument [calculus]" (p.54).[6] But the dialectic is always evident in that he also tells us that instruments are materialized theories.

Mathematical examples are used to show his understanding of the process of change in scientific thought. For example, when Euclidean geometry gave way to non-Euclidean geometry, it was not because it was contradicted or replaced, rather it was incorporated into the new way of thinking as a special case; likewise with Newtonian mechanics. He makes the case that there is not so much development of old doctrines toward new ones, but rather an envelopment of old thoughts by new (p.58).

By explaining the double approach whereby science simplifies reality and complicates reason, Bachelard believes he has shortened the path that runs from explained reality to applied thought, and it is along this path that the "pedagogics of proof" must be developed for that is "the only possible psychology of the scientific mind" (p.10)

He ponders: "Why do we always start with the opposition between a rather vague Nature and an untutored Mind, and so confuse, without further discussion, the pedagogics of initiation and the psychology of culture?" and also "How can we assert that a simple naked self can be grasped independently of its essential action in objective knowledge?" To put aside these elementary questions it will suffice to consider both the problems of science and the problems of the psychology of the scientific mind, to take objectivity as a difficult pedagogical task and not as a primitive given. Besides, "it is perhaps in scientific activity that the two aspects of the ideal of objectivity are most clear, that is, the real and social value of objectivation" (pp.10–11).[7]

"Science calls a world into being, not through some magic force, immanent in reality, but rather through a rational force, immanent in the mind. Whereas reason was, in the early days of science, formed in the image of the world, now, in modern science, the aim of mental activity is to construct a world in the image of reason" (p.13).[8]

From the above we can now see how Bachelard's interest is both in the epistemology of science and in a psychology of the scientific mind. From there it is but a short step to asking how that mind is developed. What is not always clear when he talks about pedagogy is whether it is addressed to educating the scientist or the student of science, but in the end it appears to be both. The constructivist aspect of his epistemology of science is evident from the number of times the words "construct" and "make" come up. Constructivism as a theory of pedagogy in science education is a relatively new phenomenon in that it has gained ascendance within the last two decades. It appears that Bachelard anticipated this approach but also strongly encouraged a historical approach, as we shall see.

In describing the modern scientific spirit (as opposed to mind) Bachelard is faced with a tension of acknowledging the dialectical nature of scientific epistemology historically and an attempt at some kind of synthesis or conciliation. He argues that the alleged unity of science never corresponded to a stable state, and so it is dangerous to postulate a unitary epistemology. He wonders, however, whether a conciliation (which will always be a compromise) is possible between the theoretical and experimental aspects of science. "This conciliation does not efface the dualism inscribed in the history of science, in all pedagogical development, in thinking itself" (p.15). It is for this reason that he proposes "a sort of pedagogy with (of?) Ambiguity in order to give the scientific mind (spirit?) the suppleness necessary to comprehend new doctrines" (p.15). "If one wishes to penetrate the scientific spirit in its new dialectic, one must live this dialectic psychologically, as a psychological reality, instructing oneself in the primary formation of complementary thoughts" (p.26).[9]

This is perhaps better explained in the following quote: "*It is at the moment when a concept changes its meaning that it has most meaning* it is then that it is, in all truth, an event of conceptualization. Even from the simple point of view of pedagogy—a point of view which too often fails to recognize the psychological importance—the student will better understand the value of Galileo's notion of speed if the teacher knows how to expose the Aristotelian role of speed in movement" (p.52).[10] Likewise in another example he explains that the law of reflection of light has always been known and understood. It is because of this that pedagogical development is difficult, because of the ease with the experience. It is precisely this type of "immediate given" that new scientific thought must reconstruct. In other words, experience must be problematized. "It is by fighting against primary intuition, by bringing about explanation from an experimental pluralism that one arrives at thoughts which rectify thoughts and at experiments which rectify observations" (p.75).[11]

Bachelard goes on to describe the importance of the particle and wave theory of light as an example of this process. While much of this may be reminiscent of conceptual change theory in education and the need for cognitive dissonance for such change to occur, that is not really the case. For Bachelard repeatedly insists that new ideas in science envelop the old, rather than replace them. We will come back to this observation.

In NES, Bachelard continues with an examination of the notion of determinism in light of the dialectics he has identified, and develops the notion of a non-Cartesian epistemology. It is beyond the scope of this chapter to discuss these further. Instead, let us move on to his next book, where he develops his epistemological and pedagogical ideas. What is in store is presaged by the subtitle of FES, which is "a contribution to a psychoanalysis of knowledge."

Whereas in his earlier work the notion of epistemological break is important, in FES the focus is on epistemological obstacles. While the understanding of an obstacle as a hindrance is very much in evidence in his work, there is also a positive aspect to it, in that obstacles can be the driving force for the evolution of scientific thought: "When one searches for the psychological conditions of scientific progress, one is soon convinced that *the problem of scientific knowledge must be posed in terms of obstacles*....Indeed, we know *against* previous knowledge, when we destroy knowledge that is imperfect and surmount all those obstacles to spiritualization that lie in the mind itself" (pp.13–14).

One of the things that must be fought against is an inertia that arises because of a conservative instinct whereby the mind prefers that which confirms its knowledge rather than that which contradicts it, where it prefers answers to questions. When this instinct dominates, mental growth stops, according to Bachelard. Many teachers are familiar with the situation. Like the pedagogical constructivists of today, Bachelard recognizes that scientists and students come to science with previous knowledge. To share the scientific spirit, to have a scientific mind is to be ready, throughout a thinking life, to reconstruct all of one's knowledge.

Just as in NES, the movement between individual constructivism and that of the scientific enterprise is also evident in FES. "All scientific culture must begin...with an intellectual and emotional catharsis. Then the hardest task remains: to put scientific culture on permanent alert, to replace static knowledge with knowledge that is open and dynamic, to dialectize all experimental variables, to finally give reason reasons for evolving" (pp.18–19). He goes on to tell us that these observations are at their most apparent in the teaching of science.

Scientific knowledge is opposed to everyday knowledge. It is even more opposed to opinion. One cannot base anything on opinion, it must first be destroyed. It is the first obstacle to overcome. "The scientific spirit forbids us from having an opinion on questions we do not understand, on questions we cannot clearly formulate. Above all, one must know how to raise problems. And whatever one may say, in scientific life, problems do not raise themselves. It is precisely this *sense of the problem* that is the mark of a true scientific spirit. For the scientific mind, all knowledge is an answer to a question. If there is no question, there can be no scientific knowledge. Nothing comes of itself. Nothing is given. Everything is constructed" (p.14).[12]

Bachelard tells us that the notion of epistemological obstacle can be examined in the historical development of scientific thought and in educational practice. He suggests that an epistemologist needs to approach historical facts differently than a historian and treat them as ideas which are then placed within a system of thought. He does this by presenting an evolution of scientific thought which passes through three important periods, the prescientific (the 16th, 17th, and 18th centuries), the scientific (the late 18th and the 19th century), and the era of new scientific thought, the beginning of which he fixes exactly in 1905 with the appearance of Einstein's relativity theory. These periods correspond with modes of thought that he characterizes as concrete, concrete-abstract, and abstract, respectively. For him, abstraction appears to be the epitome of modern scientific thinking.

In his later thinking, in *La Philosophie du Non,* this stance may appear to be softened a little and his description of the envelopment of previous ideas is elaborated upon in his concept of epistemological profile. It could be argued that this is a recognition of contextualized knowledge, but more than that it is a call for pluralism in philosophical culture. "An epistemological profile should always be relative to a specific concept and is applicable to a specific mind being examined at a particular stage in its culture" (PN, p. 43). He illustrates this thought in the concept of mass: mass as something heavy, bulky (naive realism); mass and the use of a balance (empirical positivism); mass as an abstract relationship between two constructs, force and acceleration (classical rationalism); mass as a variable characteristic of matter (relativity); and Dirac's negative mass (discursive rationalism). Such a profile could be cultural (as in the example) or personal (which he explores through his own understanding of energy). While he does not talk about it in these terms, in the latter case the epistemological profile could be a diagnostic tool in constructivist pedagogy (Souque, 1988). Epistemological obstacles keep us from moving through the stages. While this is generally desirable, at times working in a previous stage, depending on the context, is perfectly acceptable, as in the case of understanding mass through classical rationalism.

In FES, Bachelard takes us through the history of science to provide examples of epistemological obstacles. In many cases these historical obstacles can be compared to those encountered in pedagogical contexts. While Bachelard doesn't go quite as far, some constructivists in science education have argued that students' constructions and conceptions recapitulate historical ones. I would argue that Bachelard is closer to the mark on this issue, although I do not intend to develop that argument here. Instead, let us look at some examples that Bachelard presents from the historical perspective and link these with the pedagogical issues. It will only be possible to outline them briefly here, while Bachelard devotes at least a chapter to each obstacle.

The first obstacle he identifies is the first experience, or to be more precise, the first observation. Contrary to what is usually assumed, he makes the case that there is a rupture rather than a continuity between observation and experimentation. He is critical of the 18th century proliferation of what might be called popular science books and the public marvel at curiosities. According to him one replaces knowledge by admiration and ideas by images. This is reminiscent of the explosions one remembers from chemistry class although one has no recollection of the causes and principles being demonstrated. This kind of science teaching becomes the focus of a false interest. Furthermore, "a science which accepts images is, more than any other, a victim of metaphors. The scientific spirit must continually fight against images, against analogies, against metaphors" (p.38).

In his psychoanalytic approach to the historical periods and the obstacles he identifies, he presages postmodern approaches such as those following Lacan. For example, he shows how by observing the observer with such a perspective one understands how portraits of animals marked by a false biological hierarchy are filled with traits imposed by the unconscious reverie of the narrator. Thus the lion becomes king of the animals because the observer belongs to an order where all beings have a king.

Similarly, by examining the alchemists we come to understand what is too concrete, too intuitive, and too personal in a prescientific mentality. "An educator will therefore always have to think of detaching the observer from the object, to defend the student against the mass of feelings that is concentrated on certain phenomena too rapidly symbolized, and, in some way, *too interesting*" [13] (p.54).

In presenting the second obstacle, Bachelard tells us that nothing slowed the progress of scientific knowledge more than the false doctrine of the general. A premature and facile generalization is dangerous. While, according to Bachelard, it is important to place grand generalizations at the base of scientific culture, the difference between these and the false knowledge that they replaced is that they once raised further questions but no longer. It is in this respect that, Bachlard tells us, pedagogical stages are not homologous to historical stages for such general laws actually block thought. The historical example he gives here is the extension of the concept of coagulation, after observation of milk, blood, etc., to water. A similar obstacle is the prescientific need for unity and the rejection of duality (discussed in chapter 5 of FES, as is the obstacle of utilitarian induction, which could be considered a special case of generalization and is not discussed here).

In that example and others like it from the domain of alchemy, Bachelard warns us against any creeping-in of value. A psychoanalysis of objective knowledge has to resist all valorization. He includes in this a need to radically devalue scientific culture (p.65).

A characteristic of prescientific thought is a belief that to classify objects is to know them. Modern scientific thought is the opposite of the extension of a concept to many phenomena, instead it is to relentlessly pursue the limits of phenomena. It is objectivity, not universality, that is the goal,[14] and objectivity comes with precision and coherence of attributes, not with a collection of objects that are more or less analogous. "Knowledge that is missing precision, or better said, knowledge which is not given with its precise conditions of determination is not scientific knowledge. General knowledge is almost fatally vague knowledge" (p.72)[15]

The next obstacle that Bachelard addresses he refers to as a verbal obstacle. It is the tendency to believe that once one has given an abstract name to an object or event, one has explained it. The use of metaphors (the danger of which has been foreshadowed) also leads to obstruction. Teachers will remember students who think that the model is the thing rather than useful for illustration. Indeed, according to Bachelard, in scientific thought analogies illustrate abstract ideas; they come after the theory. In prescientific thought they come before (or are the theory).

Substantialist tendencies are another obstacle described by Bachelard. Based on his descriptions of these tendencies in prescientific thought, what envelops something has less value than that which is enveloped. For example, while the bark of a tree has important functions, it was only considered as protection for the center of the tree. In this type of thinking the substance has an interior; or better, it is the interior. Also, an immediate phenomenon is taken as a sign of a substantial property, at which point all scientific inquiry stops. The substantialist answer strangles all questions. In scientific thinking, on the other hand, each phenomenon is a moment in theoretic thought, a stage of discursive thought, a prepared result. It is produced rather than induced (p.102). The example Bachelard uses here is the substantialist intuition of Aldini with respect to electricity as a galvanic fluid and Ohm's concept of resistance. One of the symptoms of substantialist thought, according to Bachelard, is the accumulation of adjectives for the same substance so that the less precise the idea, the more one finds words to express it. Scientific progress is the reduction of the number of adjectives that belong to a substance. Likewise, it is the removal of value added to a substance when a small amount has large effects or if it takes a long time to produce.

Moving from substantialism to realism, the next obstacle is not a big step. Realism could be said to be the only innate philosophy. The dialectic between love of the real and knowledge of the real, which are almost contrary, oscillates without end, as Bachelard shows once again with historic examples.[16] As with chemistry and physics, Bachelard next takes on early biology in his examination of the animist obstacle. In this case, biological metaphors are used where one would not expect. One could, for example, talk about the fecundity of mines, or biological phenomena are used to explicate physical phenomena. Bachelard spends a chapter illustrating this with respect to the process of digestion.

What psychoanalysis would be complete without the mention of libido, the next obstacle Bachelard addresses. In this case he argues that it is impossible to think of a mystery for very long without sexualizing it. This is related to the first mystery, that of procreation, the knowledge of which parents withhold from their children. Because the libido is mysterious, everything which is mysterious arouses the libido. Immediately one loves mystery; one needs mystery. Many cultures, Bachelard tells us, have been infantilized; they lose the need to understand. Because of this need for mystery, there arise indefinite possibilities around precise laws, and this in the end forms an obstacle to abstract thought.

It is perhaps his experience with teaching adolescents that leads Bachelard to remark that if one examined the growing mind faced with a new experience, one would be surprised to find sexual thoughts. It is symptomatic, he says, that a chemical reaction where two different bodies interact becomes immediately sexualized. He gives the example of students' interpretation of the roles of acids and bases when they react. The active role is attributed to the acid and the passive one to the base. By digging a little in the unconscious (as a good psychoanalyst should), one notices that the base is feminine and the acid masculine (in French grammar—I don't know if this would hold in an English classroom). The acid then takes on a privilege of explanation based only on this attribution of activity. Bachelard also cites a text from the 17th century that takes a similar line. It is important to remember that Bachelard is critical of this way of thinking and considers it an obstacle to scientific thought. Furthermore, he is critical of teachers who do not help students overcome it: "They judge more than they teach. They do nothing to cure the anxiety that seizes every mind before the need to correct its own thought and to come out of itself to find objective truth" (p.209).

The last obstacle discussed and perhaps the most surprising (if the previous one wasn't) given our impression of modern science, is quantification. Bachelard tells us that one is mistaken if one thinks quantitative knowledge escapes the dangers of qualitative knowledge. The excess of precision in the quantitative field corresponds exactly to the excess of the picturesque in the qualitative field. One can see evidence of a nonscientific spirit even as there are pretensions to scientific objectivity. To overcome this one must be aware that the precision of measurement should constantly be referenced to the sensitivity of the method of measurement and the permanence of the object measured. "It is important to think in order to measure, not measure in order to think" (p.213). The concern with precision often leads to the posing of insignificant problems.

In this chapter on quantification Bachelard explores further the relationship between science and mathematics. He considers hostility toward mathematics a bad sign, especially when it is allied to a pretension of directly understanding scientific phenomena (p.228). He is critical of those who would like to see mathematics express a phenomenon rather than explain it. He contends that mathematics is the base of explanation in physics and that the conditions for abstract thought are henceforth inseparable from the conditions of the scientific experiment (p.231).

Throughout FES we see Bachelard raising pedagogic issues, especially with connection to overcoming the obstacles he has identified.

In the chapter on quantification and the last chapter, which is titled "Scientific Objectivity and Psychoanalysis," he deals with them much more directly. The abandonment of common sense knowledge is a difficult sacrifice (both for the scientist and the student of science). The teaching of the results of science is never a scientific education. If one does not explain the mental line of production that led to the result, one can be sure that the student will always combine the result with already familiar images. This is in line with the constructivist educators' position and those who advocate an approach to teaching science that incorporates the history and philosophy of science.[17]

He argues further, because of these tendencies of students to alternate understandings, we should have a recursive teaching of objective themes to stop subjective proliferations and to affirm an objective culture. This is especially missing at the secondary level (p. 235). But he is not here referring to a spiral curriculum. He also advocates the development of habits of discursive thought, which also develops the taste for the abstract, the need for which he returns to again and again.

The social dimension of learning and knowledge construction is also explicitly recognized. The role of the 'other' is given importance in these endeavors, once more presaging postmodern discourse that is heavily influenced by psychoanalysis. In learning, "if a child is to make progress, he must feel he is making progress *against* someone, against someone else's errors. Being right is a feeling, not just a mathematical conviction" (McAllester Jones 1991, p.80). There is an emotional component to the will to be rational. Because of this complex scientific societies are necessary, societies which would double the logical efforts with psychological efforts. He suggest that chemistry was so long in error because it was for so long a solitary culture (FES, pp.243–244).

In applying this social dimension, Bachelard provides pedagogic suggestions reminiscent of Vygotskian influences and cooperative small group learning. We should take students in a group and encourage awareness of group reason and help them to acquire an instinct toward social objectivity, "for this is an instinct that is much underestimated and in whose place we prefer to develop the opposite instinct of *originality*, failing to see that this originality which we learn from our literary studies is contrived and artificial. To put it another way, if objective science is to be really educational, then the way it is taught must be socially active....The fundamental principle of *pedagogics* of the objective attitude is this: *whoever is taught must teach*" (p.244) if we are to avoid knowledge that is fixed and dogmatic.

According to Bachelard, the psychological dimensions of being a learner and a teacher are different; the former is a kind of empiricism and the latter a kind of rationalism. Any real teaching must have this ebb and flow, and, in fact, such an alternation is critical in the history of scientific knowledge. "It is a necessity for our psychological dynamism. It is for this reason that any philosophy that confines culture to realism or nominalism only sets up the most formidable of obstacles for the development of scientific thought" (p.246).

In the conclusion to FES, Bachelard argues that not only the social context of the classroom is important to learning science, but the values he has identified need to be supported in society at large. He notes that in modern society science has not been integrated into general culture as he suggests it should be. The arguments have been that science is too difficult and too specialized. But, he counters, the more difficult a work, the more educative it is. Furthermore, only in the work of science "can one love what one destroys, can one continue the past in denying it, can one venerate one's master in contradicting him" (p.252). He finishes with a different take on lifelong learning than we are used to: "Society will be made for school, not school for society" (p.252).

While I would like to conclude with such a provocative statement, it is important to provide some concluding remarks. I believe I have made the case that Bachelard is a constructivist and a social constructivist at that. However, he is not a relativist, and while he discusses the dialectic between subjectivity and objectivity and a plurality of epistemological positions, he seems to come down on the side of objectivity and the superior position of scientific thought understood as discursive rationalism. On the other hand, he presages many ideas current in postmodern thought. More practically he has much to say to today's science educator. At the very least he deserves to be known to a larger public. I hope I have awakened the reader's curiosity and provided encouragement for others to seek out the original sources.

Endnotes

1. All biographical detail is taken from Smith (1982) and McAllester Jones (1991).

2. Of the three books which form the focus of this chapter, to my knowledge only two are translated: *The New Scientific Spirit*, translated by A. Goldhammer. Boston: Beacon Press, 1985 and *The Philosophy of No*, Translated by G.C. Waterston. New York: Orion Press, 1968. Sections of *The Formation of the Scientific Spirit (Mind?)* have been translated by McAllester Jones (1991). It is ironic that this book appears

not to have been translated since it is the one most wellknown to French students, as its sixteenth printing attests. As a result, most of the translations provided here will be my own.

3. La science est un produit de l'esprit humain, produit conforme aux lois de notre pensée et adapté au monde extérieur. Elle offre donc deux aspects, l'un subjectif, l'autre objectif, tous deux également nécessaires. This is a quote from Bouty's (1908), *La vérité scientifique*, p.7.

4. Un réalisme mathématique vient tôt ou tard corser la pensée, lui donner la permanence psychologique, dédoubler enfin l'activité spirituelle en faisant apparaître, là comme partout, le dualisme du subjectif et de l'objectif.

5. L'expérience scientifique est ainsi une raison confirmée. Ce nouvee aspect philosophique de la science prépare une rentrée du normatif dans l'expérience: la nécessité de l'expérience étant saisie par la théorie avant d'être découverte par l'observation....Le raisonnement par construction ...fait son apparence dans la Physique mathématique et dans la Physique expérimentale.

6. C'est un instrument mathématique qui crée la science physique contemporaine comme le microscope crée la biologie. Pas de connaissances nouvelles sans la maîtrise de cet instrument mathématique nouveau.

7. Pourquoi partir toujours de l'opposition entre la Nature vague et l'Esprit frustre et confondre sans discussion la pédagogie de l'initiation avec la psychologie de la culture?...Comment aussi prétendre saisir un moi simple et dépouillé, en dehors même de son action essentielle dans la connaissance objective? Pour nous désintéresser de ces questions élémentaires, il nous suffira de doubler les problèmes de la science par les problèmes de la psychologie de l'esprit scientifique, de prendre l'objectivité comme une tâche pédagogique difficile et non plus comme une donnée primitive. D'ailleurs, c'est peut-être dans l'activité scientifique qu'on voit le plus clairement le double sens de l'idéal d'objectivité, la valeur à la fois réelle et sociale de l'objectivation.

8. La science suscite un monde, non plus par une impulsion magique, immanente à la réalité, mais bien par une impulsion rationnelle, immanente à l'esprit. Après avoir formé, dans les premiers efforts de l'esprit scientifique, une raison à l'image du monde, l'activité spirituelle de la science moderne s'attache à construire un monde à l'image de la raison.

9. My moving back and forth between spirit and mind in the translation of *esprit* reflects my understanding of whether Bachelard is referring to science (spirit) or the scientist/science student (mind). The original texts

for the quotes in this paragraph are, respectively: cette conciliation n'efface pas le dualisme inscrit dans l'histoire de la science, dans tout développement pédagogique, dans la pensée même....nous proposerons une sorte de pédagogie de l'ambiguïté pour donner à la compréhension des nouveaux esprits...si l'on veut pénétrer l'esprit scientifique dans sa dialectique nouvelle, il faut vivre cette dialectique sur le plan psychologique, comme une réalité psychologique, en s'instruisant dans la formation première des pensées complémentaires.

10. C'est au moment où un concept change de sens qu'il a le plus de sens, c'est alors qu'il est, en toute vérité, un événement de la conceptualisation. Même en se plaçant au simple point de vue pédagogique—point de vue dont on méconnaît trop souvent l'importance psychologique—l'élève comprendra mieux la valeur de la notion galiléenne de vitesse si le professeur a su exposer le rôle aristotélicien de la vitesse dans le mouvement.

11. C'est en luttant contre l'intuition première, en provoquant des raisons de pluralisme expérimental, qu'on atteint à ces pensées qui rectifient des pensées et à ces expériences qui rectifient des observations.

12. Avant tout, il faut savoir poser des problèmes. Et quoi qu'on dise, dans la vie scientifique, les problèmes ne se posent pas d'eux-mêmes. C'est précisément ce sens du problème qui donne la marque du véritable esprit scientifique. Pour un esprit scientifique, toute connaissance est une réponse à une question. S'il n'y a pas eu de question, il ne peut y avoir connaissance scientifique. Rien ne va de soi. Rien n'est donné. Tout est construit.

13. Un éducateur devra donc toujours penser à détacher l'observateur de son objet, à défendre l'élève contre la masse d'affectivité qui se concentre sur certains phénomènes trop rapidement symbolisés et, en quelque manière, trop intéressants.

14. p. 71, Bachelard quotes Marcel Boll here, *Mercure de France*, 1er mai, 1929.

15. Une connaissance qui manque de précision ou, pour mieux dire, une connaissance qui n'est pas donnée avec ses conditions de détermination précise n'est pas une connaissance scientifique. Une connaissance générale est presque fatalement une connaissance vague.

16. This discussion is found in chapter 7, and much of it is written in a satiric tone. It is very dangerous here to take sentences out of context, for the meaning of a particular sentence is often the exact opposite if the whole paragraph is taken into consideration. This is often the case with Bachelard's writing, but in this chapter it is especially evident.

17. For an overview of the debates on constructivist science education see *Science & Education*, 6 (1–2), 1997. This journal is also a good source for incorporating history and philosophy into science teaching.

References

Bachelard, G. (1934). *Le Nouvel esprit scientifique*. Paris: Librairie Félix Alcan.

Bachelard, G. (1981 [1940]). *La Philosophie du non*. Paris: Quadrige/Puf.

Bachelard, G. (1996 [1938]). *La Formation de l'esprit scientifique*. Paris: Librairie philosophique J. Vrin.

Bouty, E. M. L. (1908). *La Vérité Scientifique*. Paris: Gauthier-Villars.

McAllester Jones, M. (1991). *Gaston Bachelard, Subversive Humanist*. Madison: University of Wisconsin.

Smith, R. C. (1982). *Gaston Bachelard*. Boston: Twayne.

Souque, J. P. (1988). The historical epistemology of Gaston Bachelard and its relevance to science education. *Thinking*, 6 (4), 8–13.

Chapter Three
The Simulacra of Science Education
David W. Blades

> It's the end of the world as we know it, and I feel fine.
> —R.E.M., *Eponymous*

> The absence of things from themselves, the fact that they do not take place though they appear to do so, the fact that everything withdraws behind its own appearance and is, therefore, never identical with itself, is the material illusion of the world.
> —Jean Baudrillard, *The Perfect Crime*

Introduction: Encounter with a Wolf

I was on my way home. The steady rhythm of the pedals and the familiarity of the route allowed my mind to drift freely over the events of the day. A thin blanket of snow—early this year—reflected the twilight enough to clarify the path ahead. Without warning it appeared, sauntering directly in front of me on a perpendicular route.

We both stopped in our tracks and looked at each other. Our gaze was not long or rushed. Then it moved on unhurriedly, and I regained my balance and rode away. A few minutes went by before I classified her as a wolf and wondered what she was doing in the middle of the city; even this small parcel of land called the "University Farm" was completely engineered. As I continued my ride home, I increasingly wondered—and then analyzed—the encounter. I was surprised that I had been without any fear when we met; the wolf seemingly possessed the same calm spirit.

There was nothing civilized about this accidental meeting. Words did not intervene (*A Wolf!*) and so no feelings surfaced. We just *were*, gazing for a moment that was deeply authentic and completely natural. And then she moved on and so did I; man and beast once again.

So defined, I continued my journey through the signs of modern technology: the roads, houses, street lamps, stores, and parks so carefully designed and presented. Outside the city are farms, investing in every way possible useable land to pasture or plow or forests constructed from carefully planted, genetically modified trees. Even the sky is constructed with microwaves and high–frequency radiation while overhead the atmosphere is littered with the debris of space exploration and organic compounds that erode the ozone layer. I live, as most people on this planet, in a transformed world of human creation.

Yet there are moments—very rare in my experience—when nonengineered and nonhuman constructed reality intervenes and interrupts the world without warning or preparation. The sudden appearance of the wolf, the time a bat flew into my bedroom, or the blood spilled when my son was born remind me that behind the careful order of modern society is the radical, primal existence of the natural world increasingly distant as humanity continues to mold environments for our use.

In classrooms around the world the curriculum for most children includes a subject called "science." On my ride home, I thought back to when I taught secondary school students about wolves as part of their science education. Taxonomy is one of the most common topics taught in secondary school biology classes worldwide. Wolves provide a good example of how scientific taxonomic schemes work, since students can see how wolves are related to dogs in the taxonomic plan. I remember writing the scheme on the blackboard, showing how modern science uses decreasing levels of classification to precisely identify an organism:

Kingdom — Animalia (all animals)
Phylum — Chordata (animals with notochords)
Subphylum — Vertebrata (animals with a skeleton of bone or cartilage)
Class — Mammalia (mammals)
Subclass — Eutheria (placental mammals)
Order — Carnivora (carnivores, e.g., cats, dogs, bears)
Family — Canidae (dog family)
Genus — Canis (dogs)

I knew the first question people would ask me when I arrived home and shared my encounter would be: "Are you sure it was a wolf? Perhaps what you saw was a large dog." As I placed my bike in the garage, I thought through the classification scheme one more time, recalling what I taught my students about the differences between *Canis lupus* and *Canis familiaris*, the common dog. Listing the physical differences, I felt assured that indeed, on this particular evening, I had encountered a wolf, a female, in fact. What lingered in the midst of my scientific assurance was the memory—which remains to this day—of the sudden and unexpected interruption of a wolf along my route home and the peacefulness of a moment when prior to my scientific recollections, nothing stood between us.

Baudrillard's Questioning of Science

It is science that masters the objects, but it is the objects that invest it with depth, according to an unconscious reversion, which only gives a dead and circular interrogation.
—Jean Baudrillard, *Simulacra and Simulation*

The Signs of Reality

The ambitious goal of science is to "disclose the general principles that govern the workings of the universe" (Kitcher, 1994, p.33). In science, the process for arriving at these truths is given the rather circular term, *verification*. Through verification, notes the philosopher Santayana (1983/1906), science works opposite to myth since myth "terminates in unverifiable notions that might by chance represent actual existences" (p.9), while science "terminates in concepts or laws, themselves not possible existent, but verified by recurring facts, belonging to the same experience as those from which the theory started" (p.9). For example, a scientific study of wolves would include gathering sensory input ("data") on a host of events around wolves, such as wolf migration, feeding habits, mating rituals, and the like. The theory that emerges about wolves (e.g., wolves exist in tightly knit social societies) is then verified or falsified through repeated observations of wolf behavior.

Scientific research must assume observers are sufficiently trained to ignore cultural biases or misconceptions about wolves in the attempt to gather data; that is, science believes in the possibility of approaching objectivity by minimizing the bias that humans bring to observation. The result of this belief is the generation of "facts" (linguistic or mathematical)—signifiers—that supposedly correspond directly to the reality of wolves—the signified. These facts are then used in science to modify a theory—or to further strengthen it—until the theory reaches a state where it achieves such constancy and predictive use that the theory becomes a scientific law: A fundamental truth about how the universe operates.

To arrive at such truth, science requires increasing connection between the signified and signifier through the precise use of language and mathematics. But such precision escapes experiences that seem beyond signifiers. Many who hear the howls of wolves in the night experience a creeping, primal sensation, a set of feelings so complex that words do not seem to exist to explain or even describe the event; certainly comments on the vocal tones and behavioral patterns of the night hunting of *Canis lupis* seem inadequate.

Even attempts at precision in the use of signifiers have a suspect foundation. In his essay *"Différance,"* Derrida (1982/1972) argues that "every concept is inscribed in a chain or in a system within which it refers to the other, to other concepts by means of the systematic play of differences" (p.11). The signifier "wolf" makes sense in that what it refers to is not a "dog," although there are similarities. Tracking down the similarities oddly leads us to speak of difference: Wolves generally have longer bodies, heavier shoulders, etc. With this focus on difference, Derrida demonstrates that language or any system of signifiers is thus in a state of constant flux of meanings that are never fixed and thus never precise, since *every* signifier is inscribed in a system of differences. Commenting on Derrida's essay, Madison (1990) notes that the conclusion reached from Derrida's claims is that "language is nothing but a differential system of slippage and dissemination and that meaning (as something decidable) is something that is forever deferred" (p.111).

Aside from the challenges presented by the playfulness of language, science faces a more serious difficulty in the assumption that we receive sensory input, rather than appropriate input from preexisting frames of reference. Consider the experience of wolves by the Kwakiutl people of the Pacific Northwest. In their traditional culture a separate signifier like "wolf" does not even make sense since wolves are reincarnations of the spirits of Kwakiutl hunters (Bierhorst, 1994). The best we might do in English is use "wolf-brother-honored ancestor," stringing together words in a somewhat vain attempt to understand the meaning of wolves in that society. The clearly different cosmology of the Kwakiutl from modern scientific definitions of "wolf" presents a difficulty: Which reality is real? In traditional Kwakiutl society, the taxonomic classification of wolves in biological science is nonsensical, since wolves are a kind of people. This belief is not open to verification or falsification since wolves are reincarnated human spirits, thus this aspect of seeing the world clearly lies outside the realm of modern science. We are left, then, with the uncomfortable response that the Kwakiutl way of seeing wolves is real—for that society—and science real in its particular view of the world and the societies in which it operates.

Such cultural differences challenge foundationalist assumptions of modern science and the possibility of objectivity. Kwakiutl existence maintains an *a priori* belief that spirits inhabit the world, thus when members of the society walk, for example, along a shoreline, they *appropriate* stimuli from the world in correspondence to their beliefs: They see ancestors, special visitors, hidden messages and meanings because *they seek them out* as a normal part of living in harmony with a universe full of spirits. By implication, a scientific frame of reference also provides *a priori* ways of appropriating reality, leaving questions about the validity of ways of seeing the world.

Questions concerning the principles of what counts as reality lie at the heart of the life work of French sociologist[1] Jean Baudrillard. Baudrillard states that his oeuvre focuses on "the mass effect, the mass forms that I analyze, and which, somewhere, no longer produce any difference" (1993/1983, p.45). Initially a Marxist, Baudrillard turned around Marx's notion of production as a cause of modern alienation by suggesting that the cause of the modern alienating social order today is not production but *consumption*. The events of 1968[2] led to significant changes in the thinking of many French intellectuals; during this time Baudrillard's research moved from an examination of the structural relationships between the consumer and what is consumed to the argument that the consumer was "an *effect* of the way that consumer goods circulate as meanings" (Baudrillard, quoted in Horrocks & Jevtic, 1996, p.21, italics in text). For example, while a pair of blue jeans may not vary a great deal from brand to brand, jeans presenting a label such as Tommy Hilfiger may be consumed over other brands. What captured Baudrillard's attention was how a signifier such as a label signified not only a product but also a particular meaning about the product that affected consumption. To make sense of these meanings, Baudrillard focused on the signs under which consumerism circulates. Drawing inspiration from psychoanalysis and anthropology, Baudrillard discovered that the meaning of signs exists in their 'symbolic exchange' and that what is given in a sign does not necessarily have to correspond to any physical reality. This insight moved Baudrillard into some of his most original and disturbing work.

By the early seventies Baudrillard realized that the *use* of some signs is a type of sign as well. To illustrate his point, imagine a scientist. Why is this so easy to do? Of course, there is no generic scientist, yet when asked to imagine scientists many children and adults immediately picture someone, usually a male, in a white lab coat, even though lab coats are not commonplace in the daily professional practice of many branches of science.[3] The lab coat is a signifier of science, but its use is also a sign as well. Advertisers use this signifier in generating claims about their product. This toothpaste has been, they claim, "clinically tested" and is therefore the best. How do we know? The camera in the advertisement turns to *scientists* producing report after report—presumably their objective studies of the product—vesting their claims with *authority*. How do we know the people holding the reports are scientists? They are wearing lab coats! In this way, the sign circulates as a sign itself (of authority and definite knowledge), further abstracting the signifier from the signified, which are typically very messy experiments that, to protect one's clothing if not health, require a first defense in the form of a lab coat.

Baudrillard's argument that the use of a sign can further abstract signifiers reinforces Derrida's position that signs are caught in an endless circulation of meanings and use. Baudrillard agrees with Derrida that the loss of correspondence of signifiers to reality means "the world is a game" (1993/1983, p.46) but he further comments that such a game is pathological because the real, in the sense of authentic, becomes increasingly distant in the circulation of signs to the point of no longer being present.

This extreme position could not be more opposite to the structuralist position that reality is readily available and understood through signifiers that have direct correspondence to reality (Cherryholmes, 1988; Hochberg, 1968). From a structuralist position, the analysis of relationships between signifiers leads to meaning about reality which can be shared and holds a semblance of objectivity (Cherryholmes, 1988). For example, structuralism maintains the English word "wolf" only makes sense in relation in a particular system of words where "wolf" is a sign signifying a particular set of sensations (the signified), and not, for example, those sensations that lead us to use the word "rock." That fact that it is possible to translate the signified into another system of signifiers, such as "lobo" in Portuguese, yet still refer to the same set of sensory inputs, supports the structuralist belief in the essential similarity of human experience and logic of understanding; in Shakespearean terms, "that which we call a rose by any other name would smell as sweet."[4]

At the Zoo

Baudrillard's arguments could be considered "poststructural" in his claim that, due to the growth of representation in the media and technological innovation, we no longer live in a world where signs are *obligated* to correspond to reality (Baudrillard, 1983). He points out that the use of

The Simulacra of Science Education

signs becomes a sign itself as signifiers multiply to where the sign "no longer resembles in the slightest the obliged sign of limited diffusion" (p.85). What is produced, he claims, is pure simulacra, a social game played with increasing signification more and more distant from reality. Not only are these games played with signs, suggests Baudrillard, they also "imply social rapports and social power" (p.88) arising from technology through the "presumption of an ideal counterfeit of the world" (p. 89).

In his work, *The Order of Things* (1973/1966), Foucault uses the historical development of the zoo to illustrate how this counterfeiting operates to frame perception:

> It is often said that the establishment of botanical gardens and zoological collections expressed a new curiosity about exotic plants and animals. In fact, these had already claimed men's [sic] interest for a long while. What had changed was the space in which it was possible to see them and from which it was possible to describe them. (p.131)

With the advent of zoos, argues Foucault, animals and plants became *specimens* for public display, carefully indicated with the "correct" terminology for each signified object (...on the left, children, lying in the shade is *Canis lupis,* a wolf...). He suggests that a trip to the zoo is not just a taking in of the sights and sounds but participation in a predetermined discourse on ways of seeing, hearing, smelling, and in some cases even touching a constructed reality leading to certain modes of thought about life as we gaze at the "wild" animals or "exotic" plants carefully catalogued and maintained by science. Foucault takes this a step further, however, by pointing out that not only do we receive this privileged scientific discourse, we participate and extend the discourse in our expectations of what we will see at the zoo. Historically, notes Foucault, this expectation moved from a folk naturalism to the sensational as more and more forms of life were "discovered" by Europeans traveling to distant lands. The result was the creation of zoos as a kind of theater: Visitors at the zoo expect to be informed by what they experience, but this information is anticipated as entertainment.

Baudrillard's arguments support Foucault's analysis, but considerably push the point that existing signs are counterfeits no longer obligated to reality. Zoos are often justified as a place of scientific research, a public assurance that the world is being cared for; pandas may be nearly extinct but their survival is guaranteed thanks to breeding programs at zoos. We can relax and enjoy watching pandas playing with a rubber ball. But divorced from its ecology, is a panda really a panda? Cut off from their family groupings and hunting expeditions, is what is presented as a wolf in a zoo really a wolf? Baudrillard's arguments suggest that there is nothing "wild" or "exotic" at the zoo, there is nothing present at all; the entire space is a simulacra, a constructed and managed collection of signs bearing no correspondence to reality since even the signifiers "wild" and "exotic" *are themselves signs.*

In perhaps his most significant work, *Simulacra and Simulation* (1994/1981), Baudrillard provides an example of how modern science, far from representing and investigating reality, produces signs that contribute to the distancing of human experience from reality. Consider, he advances, the case of the discovery of the mummified corpse of Ramses II. Why, ponders Baudrillard, was this mummy considered "priceless"? He suggests that Ramses

> does not signify anything for us, only the mummy is of an inestimable worth because it is what guarantees that accumulation has meaning. Our entire linear and accumulative culture collapses if we cannot stockpile the past in plain view. To this end the pharaohs must be brought out of their tomb and the mummies out of their silence. (p.10)

And so, he continues, scientists repair a mummy that was never intended to be repaired; the testimony and role of the mummy—the reality the mummy signified—is lost as the mummy becomes objectified through science and so invaded by procedures that repair the mummy which are working against the practice of the original embalmers and thus the entire spiritual significance of mummification. In our modern world, we hate to lose the mummy to the natural world; we consume the mummy instead, turning Ramses' corpse into a spectacle. No longer a signifier of a belief system, Ramses becomes a simulacrum, an "irreparable violence towards all secrets" (p.11) where everything is collected, examined, dissected, and then "preserved" and re-presented through the activity of science. As Baudrillard points out,

> mummies don't rot from worms: they die from being transplanted from a slow order of the symbolic, master over putrefaction and death, to an order of history, science, and museums, our order, which no longer masters anything, which only knows how to condemn what preceded it to decay and death and subsequently to try to revive it with science. (p.11)

The production of "Dolly" by cloning a sheep is an example where reality is presented but not preserved. In principle, Dolly is a new sheep, a marvelous if somewhat unnerving development of our technological capabilities. In an essay on cloning Baudrillard points out, however, that cloning is a kind of prosthesis of the body, an *"indefinite extension of this body by the body itself"* (1994/1981, p.98, italics in original). Rather than the creation of a new organism, he suggests, cloning destines the individual to serial propagation. The clone is not a real individual, it is "hyperreal," a creation further distanced from the reality by another media-generated spectacle for our consumption.

Everywhere, claims Baudrillard, reality is giving way to hyperreality as societies dependent on sophisticated technology are increasingly captured through media and science to consumerism. This progression to hyperreality means, according to Baudrillard (1996/1993), that we are at

> the end of labor. The end of production. The end of political economy. The end of the signifier/signified dialectic which facilitates the accumulations of knowledge and meaning, the linear syntagma of cumulative discourse. (p.440)

In his book examining the advent of the year 2000, *The Illusion of the End* (1994/1992), Baudrillard claims people in modern societies can't even speak of the end since such a discourse assumes a sense of reality and trajectory of history that for the most part slips through our hands like sand. This is because we are "no longer spectators, but actors in the performance, and actors increasingly integrated into the course of that performance" (Baudrillard, 1996/1992, p.27). In the century just past most citizens were drafted into the performance of consumerism—and most accepted their roles readily, if not greedily—only to discover that the play is a crime drama, and our role has evolved into being hostage to the message that we must consume or the machine will grind to a halt (Baudrillard, 1996/1992; 1998/1997).

Our possession in this drama deepened as the signs of our civilization pass through what Baudrillard calls the orders of simulacra (1994/1981). These successive stages move the real to hyperreality, a state where discernment is impossible in the face of what is now available to us as "reality." How can we decide what is real, asks Baudrillard, in the light of the

> communablity of the beautiful and the ugly in fashion, of the left and right in politics, of the true and false in every media passage, the usefulness and uselessness at the level of objects, nature and culture at every level of signification. (1996/1993, p.440)

"All the great humanist criteria of value," he continues, "the whole civilization of moral, aesthetic and practical judgements are effaced in our systems of images and signs. Everything becomes undecidable" (p.440).

Science is complicit in this movement, claims Baudrillard, in its quest to ensure that nothing is left to chance. The result, he argues, is an *implosion* of meaning as accident, contradiction, and rupture are systematically removed, and complexity and mystery are dissected through scientific research and technological innovation (1994/1981). Through invention, the consumer is engaged and comforted; higher and more effective fences can be built, I'm assured, to prevent wolves entering farms or disturbing my comfortable, predictable ride home. But in order for this to happen, all the remaining wolves have to be first identified and tagged—even collared—for scientific examination. Only when the wolf is no longer a wolf will the reality of safety exist.

Through the pursuit of a logic of representation and belief in the approachability of objectivity, science exposes all secrets, ironically resulting in an increasing distance from the objects of its study. Baudrillard comments that eventually the enterprise of science dispenses with objects altogether and becomes "pure," a study of reality distanced by autonomy from the mythical. In avowing the mysterious, or at least resting assured in the belief that the foundation of all causes can be known, science slips into the generation of reality for public consumption. More often than not, this "reality" is re-presented by signifiers well along the order of simulacra, a progression that is, Baudrillard argues, violent since it produces hatred by a civilization of its own foundations (1994/1981).

The Orders of Simulacra

In his review of Baudrillard's theses, Norris (1990) notes that more than any other writer Baudrillard has pushed poststructural rhetoric "as far as it will go and used it as a weapon against every last claim of truth, validity and critical reason" (p.170). The 'problem' with Baudrillard's writings, Norris confesses, is that they seem to be "amply borne out by the evidence nearest to hand" (p.171). The successive phases of the image-as-sign in the order of simulacra demonstrate Baudrillard's appeal to consider this evidence (1994/1981, p.6):

Phase One: Signs Reflect a Profound Reality
The sign below is easily recognized as the universal symbol for "no smoking." One can find this symbol in, for example, the airport of São Paulo, Brazil, where smoking is forbidden inside the airport building.

I was amused to note that almost every smoker I saw simply ignored the sign, leaving me wondering about the relationship of this signifier to its social meaning in Brazil. In the Canadian city of Victoria, the "no smoking" sign appears in *every* public building, including restaurants and bars. This has led to considerable controversy and, in some cases, outright defiance of the city laws banning smoking. In both cases the sign reflects the profound yet different realities of the social meaning of smoking in public places, linked of course to the entire discourse of the health risks associated with breathing second-hand smoke.

Phase Two: Our signs mask and denature a profound reality
Substitute a baby harp seal for the cigarette in the sign and you have the signifier used by the International Federation for Animal Welfare (IFAW), an organization that is a leading critic of the annual killing of seals on the East Coast of Canada. At the heart of this signifier lies a debate about the future of cod fishing off the coast of the Canadian province of Newfoundland. The historic practice each spring of killing and then skinning the newborn harp seals for their white fur has received international attention due to the work of agencies such as Greenpeace and the IFAW. The annual seal hunt is justified by participants as a way to control the seal population that is held partially responsible for the decimation of cod in these waters. This seems supported by scientific evidence published by the Fisheries Resource Conservation Council of Canada that seals "kill more stock north of Halifax than any other factor" (cited in Guy, 2000, p.38). Yet, the Canadian Department of Fisheries and Oceans cites contrary scientific evidence that cod are only a small part of the diet of seals.

Aside from the near impossibility of reconciling supposedly objective yet contradictory scientific studies, the human experience of the reality of seals is masked and denatured by the signs used to defend or argue against seal hunting. A full exploration of these words would fill a book, but consider a few terms: Seals are "hunted" although in every case the newborn seals lie basking in the sun on ice, available for the taking. Seals are considered a "natural resource" by all sides of the debate over the annual hunt. How does this pair of words make sense? Is something "natural" once it is considered a resource to be managed? The irony of this term, so often a part of modern vocabulary that we don't stop to consider its logic, is displayed in the word used by hunters for skinning a seal: *sculpting* (Guy, 2000). Yet while we might recoil at this term—partially because of its honesty—the discourses surrounding the reality of seals giving birth on ice flows also masks related realities such as poverty among those who worked for generations in a tradition of fishing. Further, the discourse deflects attention toward seals and not toward the problematic use of large factory fishing trawlers that literally strip the ocean of fish, a technology likely much more responsible for the decline of the cod than seals (Harris, 1998).

Phase Three: Signs Mask the Absence of a Profound Reality

I recently ventured inside a very large shopping mall to buy a new pair of running shoes, encouraged by the number of stores and hoping to find the best price. After locating a pair that interested me, I decided to compare prices at some of the other stores in the mall carrying running shoes. The price did not vary at all, even though all the stores were different in design, name, and even somewhat in the brands they carried. Perplexed, and suspecting a conspiracy of sorts, I asked a store manager about my discovery and was surprised to learn that the same company owned all the stores! Selection and competition were absent, at least with running shoes, masked by the signifiers of "different" stores. How much more is this phase of simulation becoming part of our everyday experience as corporations continue to amalgamate? How much choice exists in reality when selecting a newspaper? Television show? Food product? Despite different names, formats, and presentations, what possibilities exist in modern society to choose one product over another? In the presentation of selection in its absence and the rush to take advantage of the newest "sale," people become even further distanced from reality; a kind of fatal strategy takes over where instead of thinking in the world, the world thinks us (Baudrillard, 1998/1997).

Phase Four: Signs Have No Relation to Reality Whatsoever: They Are Their Own Pure Simulacra

The final step in simulation moves from the cover-up of absence to a system of signifiers that are their own raison d'être, a complete simulacra that, while presenting a reality, bears no relation to anything but itself. Baudrillard suggests that modernity continually moves signs from representation to simulacrum, functioning to "make reality disappear and, at the same time, to mask that disappearance" (1996/1992, p.5). It is, he suggests, the "perfect crime" (p.1): The murder of reality through the production and consumption of images in which there is nothing to see, a crime advanced and enacted, somewhat behind the scenes, on our children in the simulacra known as science education.

The Simulacra of Science Education

> From the standpoint of the child, the great waste in the school comes from his [sic] inability to utilize the experiences he gets outside the school in any complete and free way within the school itself; while, on the other hand, he is unable to apply in daily life what he is learning at school.
> —John Dewey, *The School and Society*

When I was a boy my father bought me a chemistry set. All my friends owned such kits—at least, all the boys—as we embraced as children the socially-constructed dream of becoming scientists (Blades, 1997). The instruction manual with my particular set outlined some general principles of chemistry. One phrase caught my attention: Apparently, mixing and then heating the elements sulfur and iron produces a new substance: iron sulfide. The book mentioned that this new product was a "compound" and that compounds had different properties than their constitutive elements. Now, this I had to see. At my lab bench in the basement of our home I mixed some iron with *lots* of sulfur (for good measure), scooped the mixture into a test tube, and placed it over an alcohol burner.

I was shocked by the billows of white gas coming out of the test tube! The smell of rotten eggs was so strong that my entire family and I had to evacuate the house until the smell dissipated. Returning to the experiment, I was delighted to see the silvery solid that had formed; along the traces of the test tube was the evidence I had used too much sulfur. That day, for the first time in my life, I realized that *proportions matter* when mixing substances.

My sister could have told me this years ago, having learned her chemistry through several failures in the practical art of cooking (my turn came later). The cake that didn't turn out soon led to better baking, and the sulfur dioxide gas I produced led me to learn more about how elements combine. Such experiential learning opens the world, presenting a standing invitation to ponder, explore, and understand the effects of reality.

But at the edge of this possibility many people retreat from nature as other priorities demand attention or explanations lie waiting. Something may catch our attention, such as noticing that the cut surface of an apple turns brown—this may even lead an adult to wonder why—but the question(ing) ends abruptly; there are lunches to make after all and, besides, surely some scientist has looked into this phenomenon and the answer, if anyone cares, is probably on the Internet. Levin (1988) describes this tendency toward closure of vision and possibility as the development of a "perfectly normal blindness" (p.58) to anything other than technological ways of seeing. He argues that as children, "closer to nature than to society, we began life with eyes opened by enchantment. As adults, we tend to conform to the crowd, seeing only *its* reality" (p.58, italics in original).

Parents with young children will tell you their days together are characterized by constant questions voiced from the innocence of youth: Why *is* the sky blue? Where do the geese go in the fall? But, if the plant is green so the plant can make food, how do purple plants (like plum trees) make food? Is our cat left-handed, too? Why do we need to sleep? How come other people yawn when I yawn? Why does the moon follow me? Why did the cake not turn out?

After a few years, such questions taper off. By the time children are attending secondary school, curiosity tends to be replaced with cynicism: The only questions our senior students ask is whether the topic will be on the test or, in more rare moments, how the topic has any relevance to their lives (Blades, 1992). This slow destruction of natural curiosity should not surprise us; students worldwide are captive in a system of education that substitutes exploration of the natural world for the contrived experience, understanding for memorization, and personal insight for external authority. One of the most obscene players in this drama operates under the sign of "science education" in schools: A simulacrum of the fourth order where students rarely if ever experience *science* but certainly do receive an education.

Science Education as Bulimia

Almost every jurisdiction in the world specifies, to varying degrees, what children should be learning in schools through the publication of documents usually called "curriculum guides." The discourse of curriculum documents for science education typically encourages teachers to adopt an experiential approach to science education. Consider, for example, the following overview statement concerning studies in biology at the senior secondary school level in the Canadian province of Alberta:

> Biology, *as with all sciences*, is an experimental discipline requiring creativity and imagination. Methods of inquiry characterize its study. In Biology 20–30, students further develop their ability to ask questions, investigate, and experiment; to gather, analyze and assess scientific information; and to test scientific laws and principles and their applications. In the process, students exercise their creativity and develop their critical thinking skills. (Alberta Education, 1995, p.1, italics added)

These are lovely, well-intentioned words, but distant from the lived reality of most students in the schools I have visited over the past three decades. Spend some time in most secondary school "science" classrooms and typically you find students sitting in desks, facing forward, listening to a teacher lecture. These presentations draw together a collection of "facts" from the ponderous body of information produced by modern science; often these facts are given to students in considerably more detail than required by the curriculum guides as teachers use their university lecture notes to prepare classes. The result is the accumulation of facts in the guise of science, an education that stresses, above all, the ability of students to recall the information they have been taught, usually through the imposition of external final exams (Aikenhead, 1983; Beardsley, 1992; Blades, 1997).

This common pedagogy reflects Freire's (1989) famous metaphor of education as a banking process: "education thus becomes an act of depositing, in which the students are the depositories and the teacher is the depositor" (p.58). And so students carefully record in their notebooks aspects of the canon of topics comprising secondary school science worldwide, such as the periodicity of the elements, steps on how to calculate the energy of redox reactions, the biochemical pathways of cellular respiration, or use of the universal gravitational constant—all topics memorized for the test and forgotten immediately afterwards (Blades, 1992).

But Freire does not go far enough in his indictment of educational systems. The movement of information in science education is not a benign banking but a pathology, a bulimia of information in the image of science education that is pure simulacra. The modern sickness of bulimia develops as the victim constructs, or accepts, a vision of self that is destructive. This vision enters an increasing spiral of hyperreality to the point where the prognosis can be death. Science education is a bulimic pedagogy that begins its pathology early with textbook-driven lessons in elementary schools. Almost from their first "science" lesson children learn that the canon of science contains "facts" they must assimilate. The absolute authority of these facts moves the image of science from a reflection of reality to hyperreality to simulacra in secondary schools as general concepts increasingly focus on the minutiae of the canon. For example, in their science education young children are typically taught the parts of a plant, emphasizing "correct" terminology. By the middle years of their education they can list from memory, at least before a test, the various scientific plant groups and by secondary school, students are invested in the taxonomic schemes of botany and the biosynthetic details of photosynthesis, information they had better regurgitate for the final exam if they are to obtain the grades necessary to enter university. What dies along the way is a fundamental connection to those life forms we call plants. The child does not touch, for example, a willow tree and the healing power of its bark. No one marvels at the struggle of dandelions pushing through concrete or studies the ancient use of herbs: That's not *science*. After more than a decade of education students can faithfully recite the mechanisms of gymnosperm (evergreen) reproduction and taxonomy but have no clue which parts of which evergreen can be used to make a healing tea or even how to tell the difference between a Douglas fir or a white pine—or why this might be important to know in everyday life. By graduation our children enter the terminal stages of educational bulimia: After years of learning to regurgitate information that has no relevance to their lives and no particular significance, except for the very few who become professionally involved in the discourse of modern science, their vision and curiosity die, leaving only a lingering disgust that after so many years of learning science they really have learned little that is worthwhile.

Teachers naively begin the slowly destructive bulimia of school science education by emphasizing to children how the enterprise of science deals in facts. But the existence of facts in some "true" sense requires a split of the knowing subject from objects which, Baudrillard notes, cannot be maintained ironically because modern science argues that the act of observation changes what is being observed (1996/1992). Confession of the impermanence of facts is almost never part of a "science" education, however, thus maintaining a distance in philosophy

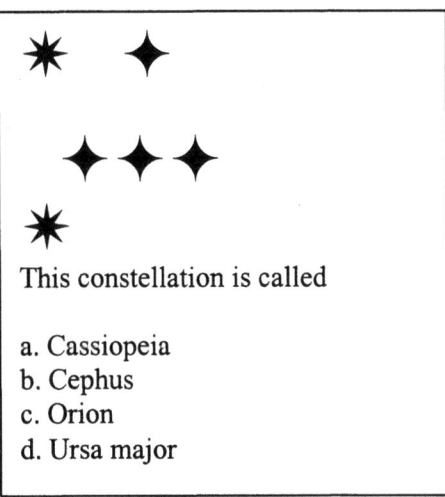

This constellation is called

a. Cassiopeia
b. Cephus
c. Orion
d. Ursa major

for what passes as science in school from the professional practice of science. Yet the inherent and liberating ambiguity of science *perspectives,* as opposed to facts, can be taught to children from a very early age. For example, the night sky is a common science education topic in most elementary schools. In an introductory unit, children learn the phases of the moon, the cause of seasons, and to identify major constellations, such as the dramatic Northern Hemisphere constellation Orion (the hunter). To ensure children have learned the topic, they are given a test at the end of the unit that may have a question such as:

This approach begins the bulimia of science education as children learn the "correct" scientific way to read the night sky. What these children do not learn is that there is nothing scientific at all about the name of constellations; most of the names used today derive from the work of the Egyptian astronomer Ptolemy (Boorstin, 1983). In those areas of the world not colonized by Ptolemy's classification scheme, the arrangement of stars called "Orion" represented a range of images, from a cow's hide, an octopus, a turtle, to part of a canoe (Staal, 1988). Some cultures, such as the Inuit of the Canadian North, did not even organize the stars into the same grouping, placing parts of the constellation into separate patterns that reflect their unique cultural stories (MacDonald, 1998).

Bound by a school schedule that operates only during daylight hours children rarely have the opportunity to actually see these stars in the night sky as part of their education. Even though a night field trip could be organized, the "facts" of the night sky tend to be presented in abstraction in the classroom. This odd removal of experience from science education is a constant feature of science education, intensifying the longer the child is in school. Speaking about her personal attempts to understand the phases of the moon, Cordeilo (1992) remarked,

> the theories about the near universe which I had held since I was a child were only founded on knowledge acquired abstractly from books and lectures and not from actual personal experience. Strange, since it's so easy to go outside and look up. (p.129)

Why are children typically not invited to look up, to touch, listen, or smell the natural world? Part of the reason is the logistics of moving children to places where such contact is possible; more often than not children live with(in) such constructed realities that authenticity is difficult to locate. But deeper than issues of convenience is the discourse of science education that provides meaning in the name of science. Students become "scientifically literate" as they acquire scientific facts, but in these attempts to master the meaning of reality, "what appears slips away" (Baudrillard, 1998/1997, p.114). Rather than being taught to challenge the authority of "facts" and to explore through experience, children begin early a careful march toward a death of their desire to know as facts are presented with authority in their science textbooks and teachers dictate notes to children that will be tested in a few weeks. This pathological, bulimic pedagogy is all the more a crime as it claims the sign of science, when there is really nothing scientific at all in the mass accumulation of facts that bear little or no meaning.

Science Education as Banal

Sometimes, if students are lucky, they break the tedium of writing to engage in so-called "experiments" to "prove" the concepts presented by the teacher. These activities, one might argue, at least provide students the opportunity to directly experience reality as well as gain insight into the professional practice of modern science. Baudrillard (1993/1983) warns, however, that "banality is the fatality of our modern world" (p.45) as the real slips to the hyperreal. The bulimia of science education reaches advanced stages as student classroom experiences of science increasingly feature nice, tidy, banal experiments that have little to do with reality.

In his critique of experiments in science education, Millar (1998) provides an excellent example of this banality in the classic middle years or secondary school biology laboratory exercise on digestion using dialysis tubing. Students fill a semi-permeable tube of plastic with a mixture of starch and glucose solutions, tie the bag shut, and leave the bag in distilled water for a length of time, often overnight. Students then test the water for the presence of starch and glucose. There is nothing experimental in this procedure; students are *expected* to discover glucose diffused into the water but not starch and are subsequently encouraged to relate this movement to the degree of permeability of the bag. Millar notes that this experiment is supposed to model the movement of substances in the process of digestion. The model, he argues, "provides no warrant whatsoever for accepting this account of digestion" (p.22) since, as he points out, the "practical work is not carried out on the real system we want the students to understand, but a model of it" (p.22).

Even the model is questionable. Digestion involves considerable mechanical movement and intestines are not passive tubes nor are intestines tied at each end! The assimilation of substances into the bloodstream is a complex process not even remotely replicated in this banal model. The experience could be more concrete if students were allowed to handle intestines of an animal provided by the local slaughterhouse or butcher. Discussions on how to test the ability of these intestines to absorb various substances allows students to enter the experience of *actual* experimentation as they contact the messy, bloody entrails of reality. Students might examine and explore how intestines, divorced from what gave them life, no longer function well, or how these intestines differ from their human counterparts, an invitation to authentic questioning instead of the banality of toying with tidy little plastic bags and following a prescriptive set of instructions.

School science "experiments" operate, argues Gough (1998), as "theatres of representations" that demonstrate Baudrillard's point that signs become simulacra as they move from masking the absence of reality to no relation to reality whatsoever. Gough lists the many ways school science "experiments" bear no relation at all to the lived world of science (pp.72–78):

1. School science is limited in scope and significance by the "individualistic, small-scale, low-tech 'bench work' to which school laboratories are suited" (p.73).
2. The valorization of the so-called scientific method in school science does not reflect the pragmatic realities scientists routinely experience.
3. The progress of science is presented as positive, tidy, and linear in schools, masking the reality that experiments are not necessarily the only way scientific knowledge develops and that there is a social-political context in the formation of scientific ideas.
4. Science education raises the exact use of science terms, such as "work," to a level of abstract correctness that bears little relation to everyday experience. A child holding a book in mid-air will experience the effort required to keep the book steady. Technically, since the book is not moving over a distance no work is completed in the scientific sense of $w = f \times d$. Aside from strict Newtonian applications, however, it is clear to the child in the everyday sense that she or he is working! Teachers, in the sign of science education, rarely resist the urge to correct the child's language and fail to make a distinction between the roles of language in science.

Gough's observations suggest that science classrooms everywhere produce a version of reality that bears no relation to the lived world of science. I was closer to science when I learned about proportions as I mixed and heated chemicals at home than when I engaged in any so-called experiments at school. Of course, there was a distinct element of danger in my home studies as there is in any experiment that by definition involves carrying out inquiries where the result is *not* known. Instead, children worldwide practice banal, safe, totally predictable experiments, where every aspect of what should turn out was already known by the teacher or available in the teachers' guide to the textbook. But the danger does not lie so much with opportunities for children to contact the world than our pathological determination to constantly invent the world for them. As Baudrillard (1996/1992) points out, the fatality of this absence is to live in the banal rather than search for the authentic; in science education this banality arises as we invite children to conduct experiments that have no experimentation, call it science, and so distance our children from anything like science: the perfect simulacra.

Science Education as Brisance[5]

Given the bulimic, banal pedagogy of science it comes as no surprise that the more students experience school science, the worse their attitudes toward science (Kass & Blades, 1992; Science Council of Canada, 1984, Yager & Penick, 1986; Yager & Yager, 1985). In the past three decades a clear erosion of public faith in scientists and antiscience sentiment (Holton, 1993), combined with the dismal experience of school science for most children, has led to an overall drop in students considering science-related careers, except in computer engineering. In response, science educators have consistently argued since the 1970s that the focus of school science on recruitment has been counterproductive and that more philosophically valid and engaging science education would consider the societal issues presented by the interactions of science, technology, society, and the environment (Duschl, 1988; Hodson, 1988). This "STS" approach to science education seeks to design educational experiences so that students can develop the "scientific literacy" necessary to act in a democracy while at the same time hoping to encourage students to consider careers in science (Aikenhead, 1980; American Association for the Advancement of Science, 1989; Blades, 1997; Zeidler, 1984).

In the name of the development of student science literacy, science education has become a media event. Shows such as "3-2-1 Contact" or the popular "Bill Nye the Science Guy" appear to directly challenge the banal image of science by moving in the precisely opposite direction: Science is represented as a fast-paced, exciting investigation of the natural world, an adventure designed to attract children to science. Bill Nye, for example, is a former comedian who "bungee jumps, SCUBA dives, parachutes, and does all sorts of cool stuff with science" (Disney.com, 2000, p.1). Dressed in the traditional signs of science—white lab coat and geeky bow tie—Nye enthusiastically moves audiences through a series of quick film clips, amusing anecdotes, science demonstrations, and features of "way cool" scientists in an effort to amuse, captivate, and entertain. Recent additions to the Bill Nye approach include a series of Internet web sites that contain advice to teachers, ideas of experiments children can do in their homes, and features from past episodes of the television show.

Sponsored by the National Science Foundation and produced by the Walt Disney Company for public television, this show has won more than a dozen daytime Emmy awards. Described as an "ambassador for science" (Nye, 2000), his appeal to children that science is fun and interesting has led to a series of honors, including participation in the President of the United States' round table forum on education, appointment to the U.S. Department of Energy Task Force on Education, presentation to the House Committee on Science Education in the

Congress of the United States, and Guest of Honour at the dedication of Cornell University's memorial to the late astronomer Carl Sagan.

It is hard to watch *Bill Nye the Science Guy* closely. A certain relaxation takes over; science becomes fun again, or appears to be fun. A rapid succession of facts slide between shots of children trying an idea suggested to them by a teacher; suddenly Bill is stomping through a marsh talking about wetlands; I laugh out loud. Everyone is cool: the teacher, the children, scientists and, of course, Bill the most. Without a doubt, this is entertainment, an excellent example of what Baudrillard (1983) describes as a "hallucinatory resemblance of the real with itself" (p.142) in the best tradition of Disney productions. But despite the sometimes frenetic pace of the presentation, the glitzy and slick presentation, Bill Nye the Science Guy still retains the tradition of predetermined "experiments" and the authority of science "facts" in science education; it is in the final analysis a simulacrum of science. At no point do students hear about the difficulty of scientific research, the long tedious hours of hunting an idea that more often than not ends in a blind alley. Students are never seen thinking of an experiment themselves or engaged in the social action considered essential in an STS approach. The show is brisance, a blinding explosion of images that are hyperreal, distant from any reflection of the profound reality of science and, like an explosion, leaving only images and traces of a presence once the smoke has cleared.

Science education has long provided entertainment for television. Currently, there is such a current demand for science shows that entire channels are exclusively devoted to science. Twenty-four hours a day anyone with access to cable television in Canada or the United States can marvel at the sexual potency of lemmings, learn about NASA missions to planets and moons, be astounded at the destructive power of tornadoes or volcanoes (you can choose), watch wild dogs engage in social interactions on the African savanna, or be astounded at the power of the Great White Shark, all with a glorious sense of detachment, safe in the comfort of their homes.

I admit to enjoying these shows, but as Baudrillard (1993/1987) observes, there is something more than peculiar about these images:

> If they fascinate us so much it is not because they are sites of the production of meaning and representation—this would not be new—it is on the contrary because they are sites of the *disappearance* of meaning and representation, sites in which we are caught quite apart from any judgement of reality, thus sites of a

fatal strategy of denigration of the real and the reality principle. (p.194, italics in text)

What is presented in the sign of science is the hyperreal, a trip to the zoo from the comfort of our chairs. What is represented is not present in any sense; how can the lives of mountain gorillas even hope to be present in any authentic way when the lights, sounds, smells, and talking of camera crews changes their environment? Through television we become voyeurs of the natural world as science—from space shuttle broadcasts to the glimpses of the ocean floor—is offered for consumption, a small brisance every thirty minutes. The collections of facts that characterize these shows of science may lead to concern about the loss of Amazon rainforest, but only for a few minutes and certainly not enough to consider the deforestation in our backyards; at the commercial it's time to get something to eat.

The sad consequence of this indifference, suggests Baudrillard (1996/1992), is that "we no longer know what to do with the real world. We can no longer see any need whatever for this residue, which has become an encumbrance" (p.42). Motorists traveling along freeways through the national parks kill—and are sometimes killed—as nature strolls across the roadways at night. It's an irritating problem and hopefully overhead bridges for creatures, such as wolves, will allow them to migrate over highways without bothering us or contacting us at all. Our science education is thus an education about our relation to the natural world, the reminder that through science and technology humanity can solve these uninvited intrusions, encouraging an acquiescence to ways of thinking that began with an education that taught us not to be bothered by anything difficult or messy like reality: just learn the facts for the test next week.

Is an Authentic Science Education Possible?

Now, is it the function of education merely to help you to conform to the pattern of this rotten social order, or is it to give you freedom—complete freedom to grow and create a different society, a new world?
—Krishnamurti, *Think on These Things*

By the mid-1980s, many jurisdictions around the world were striving to reform science education toward an STS approach (Bybee & Mau, 1986). The depressing situation is, however, that these well-intentioned reform efforts have "done more to stabilize an obsolete curriculum than to provide insight and guidance for realizing a new vision of science teaching" (Hurd, 1991, p.35). Our vision is missing because we are trying to reform what isn't there: Schools do not engage in science education but simulacra of science education. In the lived curriculum of classrooms, what operates under the sign of science education continues to be a content-driven, bulimic pedagogy driven by standardized testing that caters to the intellectual elite bound for postsecondary study.

In *Simulacra and Simulation* (1981/1994) Baudrillard argues that the difficulty of change is due to the way a simulacrum can last indefinitely because it

> is nothing but the object of a social *demand*, and thus as the object of the law of supply and demand, it is no longer subject to violence and death. Completely purged of a *political* dimension, it, like any other commodity, is dependent on mass production and consumption. (p.26, emphasis in text)

His point suggests that the demand for television shows enticing children to be entertained by science will be counterproductive to realizing science education in the lives of children, assuming that is even remotely the intention. What these shows accomplish is the entrenchment of the simulacra of science education—and the generation of revenues—at least until boredom overtakes the audience and the show is cancelled. But the mass production of what is called "science" in our schools through centralized school systems and the relentless march of standardized testing ensure that sooner or later someone will reinvest in the production of television broadcasts that feature a more entertaining science education than can be accomplished in classrooms typically lacking the technological wizardry. In the end, however, science becomes in classrooms or on TV a simulacrum produced by social demand for an education purged of the political act necessary in the actual study of the world.

In its attempt to understand the natural world, science is an "untidy, unpredictable activity that requires each scientist to devise her or his own course of action" (Hodson, 1998, p.19). The dynamic nature of choosing a course of action in scientific inquiry is always political in origin because it is embedded in the culture from which these acts arise and political in direction since discovery presents a call to further action. Scientists know this only too well; the discovery that a certain chemical blocks receptors to pain in human brains reflects a cultural desire to provide chemicals to relieve pain, rather than other avenues of pain relief (such as acupuncture, massage therapy, etc.). Yet the discovery of this chemical means the scientist must check for the *other* effects of introducing this chemical into the human body and the consequences of ignoring warning signs masked by the use of the chemical. *Every scientific discovery is linked to the context that led to the discovery and the responsibility revealed by the information produced.*

Baudrillard's work demands that we include in the political activity of science the role of science in the production of simulacra. The generation of images and signs in science in the guise of privileged "truth" produces accomplished and excessive meanings for consumption. Such abandonment of humility functions to make reality disappear and, at the same time, to mask that disappearance" (1996/1992, p.5): the production of simulacra. This process has always been a part of science, but as handmaiden to recent technological innovations, science has dramatically increased the production of simulacra to include humanity itself. The cloning of mammals, "advances" in genetic engineering, and development of humanoid robots challenge conceptions of what it means to be human in the world, leaving us wondering where these innovations are heading (Blades, 1999). This is the final chapter of simulacra; a destining developed through science and effected through consumption. It's discouraging to know that as soon as androids become commercially available—likely within the next few decades—people will purchase them, regardless of the implications of this invention for human existence. The ethical issues surrounding cloning may be the subject of classroom discussions, but in the end people will elect to use cloned products, such as organs, if they can afford to. Societies may hate where science seems to be taking us, but actually the journey is not in the hands of science but consumers dedicated to embrace the next sign of a successful life in the form of the newest, best, and ready product—until finally humanity itself is produced as our merger with our technology becomes complete.

Toward a Hopeful Science Education

It is an understatement that Baudrillard paints a discouraging picture of the modern situation. By taking poststructuralist arguments to their logical end in the area of representation and linking these points to keen observations about consumption in modern, technological societies, he leaves little room for hope that the destructive course of these societies can change. Certainly, the present display and activity of science education in our schools make the situation even worse.

Granting assent to the major points made by Baudrillard, his poststructuralist stance falls into the classic relativist paradox, expressed well by Norris (1990): "If he [Baudrillard] succeeds in undermining all appeals to truth, validity, or rational warrant, then there can be no grounds for counting him right on this or any other question" (Norris, p.184). In particular, just because simulacra appear, does this mean that *every* human experience is ruled by codes to the point where all experience is inauthentic? There is a certain absurdity in telling someone having a baby they are part of the simulacra of childbirth; the reason we call this process "labor" in English is still obvious to the person giving birth. In a more abstract sense, persons subject to derision or persecution because they have a certain skin color or embrace a religious viewpoint hear the comments, feel the exclusion and humiliation and authentically experience persecution, even if we agree that the word "community" is a simulacrum of their neighborhood. Human experiences such as suffering, joy, and love may be the product, in many cases, of a consumption of signs distant from reality; children who "suffer" because they do not have the "right" pair of jeans are an example. But we should not place this example with the victims of war who suffer through torture, starvation, or extermination. We may agree with Baudrillard that the *meaning* of these events quickly becomes part of a spectacle for consumption, even a simulacra, but this does not mean the event lacks authenticity for those who lived through it or live in its shadow.

The fact that Baudrillard's arguments *are* compelling presents another paradox in poststructuralist thought. While words as signs may circulate as meaning, nevertheless there must be some belief that enough *shared* meaning exists that communication is possible. Baudrillard publishes books, a very hopeful act given his stance on how the consumption of meaning leads to the production of simulacra. I suppose he would wish his writings somehow exempt from the processes he articulates; in a sense they display immunity in that his arguments can be understood. The words on the pages published under his name are not scrambled signs or a collection of meaningless symbols but, presumably, carefully chosen representatives with the hope of communicating his beliefs and ideas. The ability to accomplish this intention, with varying degrees of success, limits the poststructural argument to the recognition that, among humans, communication is possible.

Almost any animal backed into a corner will fight for its life; Baudrillard's communications have a similar effect: Do we disappear or find hope? In an interview (1993/1983) he admits to writing as an act of personal challenge and a way of defying the creeping production of reality, an invitation through the publication of his work to join in this resistance. In the same interview, he ponders the question of hope from the challenge of his analysis of modern society:

> Is the world irredeemably lost? I don't know. It is so functionalized that everything you might call game, illusion, even language itself, risks remaining caught up in it. However, let's not indulge in complacent catastrophism; things develop by themselves, perhaps, and to apply to them pseudo-moral ideas of humanist deontology—that won't change anything at all. And then, there will always be a way of playing with the systems. (p.48)

At the end of his analysis, then, Baudrillard's claim that he is not a prophet of doom (p.43) is valid since he hints that the authentic, while typically doomed to obscurity through the appropriations of science, is nevertheless still available and that, as demonstrated by his writing, communication is possible. In this way, Baudrillard offers a penetrating and largely, I believe, accurate diagnosis of modernity, but his writings also confirm that the prognosis is anything but certain.

The education of children begins with communication by the child's caregivers. Even though this communication socializes the child to particular realities, communication does not necessarily mean the inevitable distancing from authentically experiencing the world. The range of possibilities communicated with a child may be narrowed by culture, but the challenges presented by Baudrillard's writing also demonstrate that communication can *open* vision. Baudrillard hints at this hope at the end of one of his interviews when he admits that the only strategy against the relentless production of simulacra may be the discovery of the blind, tangential points of reversion in systems of productions (1998/1997, p.115). This also explains why Baudrillard admits to an interest in myth (1993/1983), for in the edifice of modern science myth presents a critical voice by claiming that other, legitimate ways of approaching reality exist besides those represented by modern science.

These points of rebellion appear through a fundamental questioning of the social order, precisely as demonstrated by poststructuralist thought. Questioning the realities that are given suddenly reveals the fragility of the systems that both capture and produce meaning. For example, upon the thoughtful examination of a label on a pair of jeans a teenager taught to question might suddenly realize that the label produces meaning that a person is cool only when s/he wears a particular brand of jeans. "How did this happen?" the teen asks. "Does the price reflect some reality about the quality of the product, or is this all a game?" The teen may decide to play the game or not, but from the moment such questioning appears this person is no longer completely caught in a system of thought and the road to freedom becomes a bit less of an uphill climb.

Educating children to fundamentally interrogate what is given as reality is actually not difficult, since children begin their lives with wonder, awe, and deep curiosity. Creating an environment that encourages this questioning to continue is the challenge. The child's home can promote wonder by cherishing questions and providing opportunities for the child to explore personal interests. One of my fondest memories as a very young child is exploring a tidal pool with my mother. The incredible array of living things caught at low tide captivated me even more as my mother revealed that hiding under the small kelp were tiny rock cod and the way tiny sea anemones try to capture your finger when you touch them. We spent what seemed to me hours marveling at the miniature ecosystem; as an adult I wonder now if my childish rapture revealed again to my mother a world she had forgotten in the business of living.

From their first day in school, however, children are systematically removed from their curiosity and wonder as they learn the authority of the text. The regularity of the mimeographed sheet, the textbook reading, and the banal "experiment" continue to dominate what passes as science education. I'm happy to report, however, that I know teachers who, when a question arises in class, continue to find joy in not knowing everything, challenging the students with the important words: "Let's find out!" In my experience, however, these teachers are rare and seem to become more so as children progress through the system.

The sheer amount of facts teachers feel they must present to children and the prevalence of achievement testing at various grade levels serve to minimize the opportunities for teachers to allow a child's question to enter the science classroom. The resultant irony is a science education without questions, except those already provided! There appears to be little room to break from this tradition, however, as Baudrillard observes, there are always ways to play with the system itself, suggesting that while entrapped, the trap itself might provide a means of escape. This movement will not be easy, but Baudrillard's work suggests a way of proceeding if we consider a pedagogy that intentionally reverses the procession of simulacra. This reversal is not intended as a prescriptive sequence of steps but points of rupture that can occur at any point in a child's education as science education moves to authenticity by making the existing simulacra the object of play.

Just Say No to Scientism

The fourth and final step in Baudrillard's succession of the image (1994/1981) is the movement from absence to simulacrum to a point where the image has no relation to reality whatsoever (p.6). To reverse this situation, children need to be invited in their science education to engage in opportunities to investigate their questions. Even though young children are naturally curious, this is not an easy step for the content demands of established curriculum leave little room for children to ask questions, let alone find answers. However, the pedagogical approach of science education can become more authentic if children are invited to *first* become explorers, following the questions that inevitably emerge through study and experimentation and *then* having exposure to the canon of scientific knowledge. This is the complete reversal of the traditional approach of classroom science education that establishes the authority of the scientific knowledge and then seeks to prove this authority by a contrived "experiment." For example, the topic of weather is a common part of the curriculum of children attending elementary and middle years schools. Teachers could initiate this topic by inviting children to regularly go outside the school and note weather conditions, gathering a set of records over time. There is no reason why the children could not suggest which records they might keep in attempting to meet the challenge of discovering a pattern in the type of weather and conditions the previous day. Only after the children have a body of experience with weather would the teacher invite the children to examine the scientific knowledge associated with weather.

Inviting children to first explore and experiment invariably introduces complex questions about the science of weather forecasting. Children may ask about the accuracy of a relative who senses weather changes "in her bones." The surprising correlation between weather and animal behavior may come up as a question. Teachers who have tried an experiential approach to teaching about weather often find a more difficult question emerging: Why are the published forecasts so often inaccurate? This questioning allows teachers the opportunity to move science education from simulacra by revealing to children the limits of scientific inquiry and the radical thought that scientific knowledge lacks a certain precision, which is why weather forecasts are usually given in percents: Cloud and wind patterns *often* bring a certain type of weather, but not always—it's a complex business. The difficulty of being accurate despite years of experience suddenly reveals to children that even in the arena of making sense of the natural world science is a limited discourse.

This runs counter to the prevalent scientism in such locations as commercial advertising, nature shows, and zoos. Scientists may assure us that dolphins held in captivity are psychologically healthy. Through their science education children can be taught to question this authority. How can anyone know for certain the psychological health of a dolphin? What evidence is used to make this claim? To what extent are the opinions of the scientists precise? How are scientists, hired by the zoo or a shopping mall,[6] constrained by their employment? What political situations, such as public attitudes about dolphins, affect the opinions provided by the scientist? By adopting a pedagogy where children *continue* to naturally ask questions by exploring the natural world before conversing with scientific knowledge, science is allowed a more humble position in the social order, reversing the scientism in school science education that privileges science knowledge to the point of producing simulacra.

Recovering Myth

The third step in Baudrillard's progression occurs as images develop from masking a profound reality to absence of the reality in the sign. School science education could restore presence from absence by examining the ways reality is masked in the name of science to the exclusion of other ways of knowing. Science teachers can further open conversations about the limits and values of scientific knowledge in understanding reality by respectfully and thoughtfully inviting myth as part of the restoration of authenticity to science education. In this invitation, myth is not appropriated as a counterpoint to science but as a complement to the human experience of reality. For example, the Abenaki people of the Northeast Woodlands (Caduto & Bruchac, 1989, pp.67–71) relate a story about wind. In their story the hunter Gluscabi decided to go duck hunting on a nearby bay, but every time he ventured out on his canoe, the prevailing winds pushed his canoe back to shore. Prying from his grandmother the location of the Wind Eagle, Gluscabi eventually finds the great Eagle, slowly flapping his wings to create wind. Gluscabi tricks the Eagle into moving locations until the Eagle falls into a crevice and the wind ceases. Gluscabi returns to hunt ducks but discovers that without wind the world becomes a terrible place: he begins to stink from sweat; the water turns ugly from pollution, and

The Simulacra of Science Education 87

breathing becomes nearly impossible. His hunting ruined, Gluscabi returns to his grandmother, who tells him after he confesses his deed:

> Will you never learn? Tabaldak, The Owner, set the Wind Eagle on that mountain to make the wind because we need the wind. The wind keeps the air cool and clean. The wind brings the clouds which gives us rain to wash the earth. The wind moves the waters and keeps them fresh and sweet. Without the wind, life would not be good for us, for our children or our children's children. (p.70)

Gluscabi, understanding the seriousness of his actions, returns to Wind Eagle to its proper location, and the wind returns.[7]

We cannot travel, like Gluscabi, to see the Wind Eagle for ourselves; indeed, the idea is mildly offensive and misses entirely the point of the story: Humanity should not tamper with the forces that drive weather systems, such as wind. The story is mythical but not without a fundamental connection to reality. Is it scientific? In the sense of being open to experiments that can be replicated it is not, but in uncovering an essential insight into the relationship between humanity and the world, the myth belongs in the discourse of science.

The deep insights provided by the myths of ancient peoples and the collective wisdom of societies—such as my English grandfather's knowledge of herbs—can be part of the conversation on the relationship of humankind to reality. Modern science is a natural part of this conversation but not the only participant; instead of delegating myth to the impoverished status of superstition, science education can invite these fundamental and important stories to the classroom to help children realize that there are many facets to representations of reality, each with an insight to contribute to understanding.

This does not mean that superstition does not exist or that some perspectives, even if well entrenched in a culture, are not problematic. Science education can also provide opportunities for students to investigate beliefs that *are* open to inquiry. An example is the superstition that parts of an animal can be used to prepare potions that empower humans, such as the belief that the horn of a rhinoceros makes a potent aphrodisiac. Aside from the disastrous results of this belief to the population of rhinoceroses around the world, it is possible to investigate this idea through a double-blind placebo experiment, revealing the superstition as an idea ineffective on humans and harmful to another species. Incorporating myth into science education is not the abandonment of modern science either, but the beginning of a conversation on how humankind comes to understand reality from a sociocultural perspective, including those aspects, such as dangerous superstitions, that need to change. By maintaining constant questioning that begins with exploration of the world and including myths, science education can rediscover with children the nature of science and the role and limitations of science in finding a hopeful future.

Contacting the World Responsibly

The final step in reversing the procession of simulacra is the most difficult: Moving away from signs that reflect a profound reality to a direct contact with reality. Since language is a cultural sign system that immediately interprets reality, a certain magic is required to find the world without culture and language immediately giving a sign of our experience.

These moments of magic are rare but possible in our lives as interruptions of the world amidst the comfort and safety of constructed realities: A wolf on the way home. There is a certain spirituality to these moments, not in the sense of organized systems of religion but in the possibility of a fundamental contact with the world where, for a moment at least, we realize our mutualism with other forms of life and the planet itself. Baudrillard suggests reality at these moments is "profound" in its authenticity and I agree, but I do not share his pessimism that our social order has developed simulacra to the point where reality has disappeared from us forever; at least, I believe and hope not yet.

With our children, we can venture forth into the world in deeply spiritual ways, contacting the primal reality we call "Nature." At this stage science education recovers its ontological dimension by allowing spirituality to become a normal part of science education. Such an addition is warranted, argues the philosopher William James, because modern science is not empirical *enough*. Challenging a restricted view of science, James suggests that science should not be limited by sensory input but should include studies of memory and the sense that "something is there" underlying sensation (James, 1962/1948). His writings consistently point out that religious and mystical experiences as well as feelings of worth, responsibility, and love are empirical since these are part of the human experience.

The essential spiritual aspect of science is like a well-kept secret in schools. Yet consider the words used and capitalized by the astronomer Carl Sagan (1995) in his last book before his untimely death:

The Simulacra of Science Education

> In its encounter with Nature, science invariably elicits a sense of reverence and awe. The very act of understanding is a celebration of joining, merging even if on a very modest scale, with the magnificence of the Cosmos. (p.29)

Students watching plant cells under the microscope move their contents in response to light are often astounded by the complexity of plants; standing in the presence of Northern Lights in the night sky, people are often moved to silence; seeing salmon fight to reproduce before dying in a steam can brings tears of compassion to witnesses. It is human nature to contact the world and to be moved, surprised, and challenged by it.

Every science class could be an opportunity to restore the sense of reverence that comes from contacting the natural world, but children must also understand that contact also changes Nature in profound and often harmful ways. It is not enough, then, to bring children to a forest or stream or night sky. To marvel begins spirituality but quickly devolves into entertainment—moving along the production of simulacra—without responsibility. The science education of our children is only able to approach authenticity as children realize that every contact with Nature changes reality in some way and thus demands a carefulness and thoughtfulness as we live within the world.

The word 'responsibility' is derived from Latin *respondere*, meaning 'to promise in return' or 'to give a reply.' How might the next generation give a reply to Baudrillard's indictment of modern society? In light of technological innovations that challenge what it means to be human, degradation of the environment and generation of simulacra, the world our children inherit is complex, difficult, and threatening. But the next generation does not have to face the future alone. Through their science education that deconstructs scientism, embraces myth, and encourages responsibility, we can help form a hopeful vision of the future.

Finding this hope involves providing opportunity for children to *act* on their knowledge in ways that counter the growing simulacra that move us away from contacting reality. Students might, for example, challenge the sign that claims an area deforested is a "managed forest." The area may be managed, but how is it still a forest when clear-cutting and reforestation reduce the numbers of species of trees? In their unit on plant life, students might compare the differences in "Round-up Ready Canola"—a genetically modified plant—to other forms of canola. The difficult questions that emerge can be placed in the political arena, such as why Canadian consumers are not given choice over which type of canola product is used in their food supply. We may be formed by the signs of modern society, but we are not helpless in this formation since we do not have to play the games they present. Teaching children how to accomplish this destruction of the effects and role of signs must be a primary focus of science education if humanity is to find a way past the realities offered for our consumption.

This accomplishment, which I believe is possible, provides a response to the charges Baudrillard reads against the "perfect crime" that is modernity. He has read well the development of the signs of modernity, but his discouragement is premature. Children are in schools—the largest captive audience in human history—presenting the opportunity of a lifetime to find ways of developing an authentic science education. This optimism is hopeful because, without warning and despite the careful construction of reality offered to us, a wolf crosses our path, inviting us home.

Endnotes

1. In an interview, Baudrillard claimed he was not a sociologist since he lacked the necessary training. Given the tone and topic of his work and his appointment to a university department of sociology, I have chosen to use this term and not the more vague, "theorist" Baudrillard prefers; despite his claims on not being formally trained as a sociologist, his astute observations of society in the twentieth century form a serious and important body of sociological study.
2. Such as the Paris student revolts, FLQ crisis in Canada, assassination of Robert Kennedy and Martin Luther King, Jr., and the Russian invasion of Czechoslovakia.
3. A friend asked her students in the sixth grade to draw a scientist. Of her 26 students, 25 drew a male in a lab coat with crazy hair doing some outrageous experiment. Most were copies of the portrayal of scientists on popular television shows. In their study of children's views of scientists, Chiang and Guo (1996) found similar stereotypes. It is hard to understand if these views are the actual views held by children or the children having fun with the image of the scientist. Regardless, almost exclusively scientists are identified as wearing lab coats.
4. *Romeo and Juliet*, Act II, Scene 2.
5. The shattering power of high explosives.
6. Shows featuring dolphins are an "attraction" at the so-called world's largest indoor shopping mall, the West Edmonton Mall in Edmonton, Alberta, Canada.
7. My brief summary fails to capture the intricacies or humor that make this a thoughtful and delightful story; I apologize for the poor telling.

References

Aikenhead, G. (1980). *Science in social issues: Implications for teaching*. Ottawa: Science Council of Canada.

Aikenhead, G. (1983). Teaching science today, relevant to 2001 A.D. *Teacher Education* 23: 58–75.

Alberta Education (1995). *Biology 20–30*. Edmonton, Alberta: Author.

American Association for the Advancement of Science. (1989). *Science for all Americans: Project 2061*. Washington, D.C.: Author.

Baudrillard, J. (1983). *Simulations* (P. Foss, P. Patton, and P. Beitchman, Trans.). New York: Semiotext(e). Original work published in 1983.

Baudrillard, J. (1993). The power of reversibility that exists in the fatal. In M. Gane (Ed.), *Baudrillard live*, pp. 43–49. New York: Routledge. Original work published 1983.

Baudrillard, J. (1993). The evil demon of images and the precession of simulacra. In T. Docherty (Ed.), *Postmodernism: A reader* (pp. 194–199). New York: Columbia. Original work published 1987.

Baudrillard, J. (1994). *Simulacra and simulation* (S, F. Glaser, Trans.). Ann Arbor, Michigan: University of Michigan. Original work published 1981.

Baudrillard, J. (1994). *The illusion of the end* (C. Turner, Trans.). New York: Verso. Original work published 1992.

Baudrillard, J. (1996). Symbolic exchange and death. In L. Cahoone (Ed.), *From modernism to postmodernism: An anthology* (pp.437–460). Oxford: Blackwell. Original work published 1993.

Baudrillard, J. (1996). *The perfect crime* (C. Turner, Trans.). Palo Alto, CA: Stanford. Original work published 1992.

Baudrillard, J. (1998). *Paroxysm* (C. Turner, Trans.). New York: Verso. Original work published 1997.

Beardsley, T. (1992). Teaching real science. *Scientific American* 267 (4): 98–108.

Bierhorst, J. (1994). *The way of the Earth: Native Americans and the environment*. New York: William Morrow.

Blades, D. W. (1992). Conversations from the margins: Grade 12 students' perceptions of their senior high school science program and possibilities for change. *Alberta Science Education Journal* 25 (1): 8–18.

Blades, D. W. (1997). *Procedures of power & curriculum change*. New York: Peter Lang.

Blades, D. W. (1999). Habilidades básicas para o próximo século: desenvolvendo a razão, a revolta e a responsabilidade dos estudantes

(Basic skills for the next century: Developing students' reason, rebellion, and responsibility). In Luiz Heron da Silva (Organizador), *Século XXI: Qual conhecimento? Qual currículo?* (pp.33–61). Petrópolis, RJ, Brasil: Editora Vozes.

Boorstin, D. J. (1983). *The discoverers.* New York: Random House.

Bybee, R., & Mau, T. (1986). Science and technology-related global problems: An international survey of science educators. *Journal of Research in Science Teaching*, 23, 599–618.

Caduto, M. J., & Bruchac, J. (1989). *Keepers of the Earth.* Saskatoon, Saskatchewan: Fifth House.

Cherryholmes, C. H. (1988). *Power and criticism: Poststructural investigations in education.* New York: Teachers College.

Chiang, C-L., & Guo, C-J. (April, 1996). *A study of the images of the scientist for elementary school children.* Paper presented at the Annual meeting of the National Association for Research in Science Teaching, St. Louis, Missouri.

Cordeilo, P. (1992). *Whole learning: Whole language and context in the upper elementary grades.* New York: Owen Publishers.

Derrida, J. (1982). *The margins of philosophy* (A. Bass, Trans.). Chicago: University of Chicago. Original work published in 1972.

Dewey, J. (1990). *The school and society.* Chicago: University of Chicago Press. Original work published in 1900.

Disney.com (2000). *About BNSG.* Available: http://disney.go.com/DisneyTelevision/BillNye

Duschl, R. A. (1988). Abandoning the scientistic legacy of science education. *Science Education*, 72 (1), 51–62.

Foucault, M. (1973). *The order of things.* New York: Random House. Original work published 1966.

Freire, P. (1989). *The pedagogy of the oppressed* (M. B. Ramos, Trans.). New York: Continuum.

Gough, N. (1998). 'If this were played upon a stage': School laboratory work as a theatre of representation. In J. Wellington (Ed.), *Practical work in school science: Which way now?* (pp. 69–89). New York: Routledge.

Guy, R. (2000). Seal wars. *Canadian Geographic* 120 (2), 37–48.

Harris, M. (1998). *Lament for an ocean.* Toronto: McClelland & Stewart.

Hochberg, J. E. (1968). Perception: Introduction. In D. L. Sills (Ed.), *International encyclopedia of the social sciences* (Volume 11, pp.527–535). New York: The Macmillan Company.

Hodson, D. (1988). Toward a philosophically more valid science curriculum. *Science Education*, 72 (1), 19–40.
Hodson, D. (1998). *Teaching and learning science*. Buckingham: Open University.
Holton, G. (1993). *Science and anti-science*. London: Harvard University Press.
Horrocks, C., & Jevtic, Z. (1996). *Baudrillard for beginners*. Cambridge, England: Icon Books.
Hurd, P. D. (1991). Why we must transform science education. *Educational Leadership*, 49 (2), 33–35.
James, W. (1962). Pragmatism's conceptions of truth. In W. Barrett & H. D. Aiken (Eds.), *Philosophy in the twentieth century* (Volume 1, pp. 193–207). New York: Random House. Original work published 1948.
Kass, H., & Blades, D. W. (1992). *A post-modern perspective on the marginalization of student voices in science education curriculum–discourse*. Paper presented at the 1992 Fourth Annual Conference, International Consortium for Research in Science and Mathematics Education, San Juan, Puerto Rico, February 3–4.
Kitcher, P. (1994). *Abusing science*. London: MIT.
Krishnamurti, (1964). *Think on these things*. New York: Harper & Row.
Levin, D. M. (1988). *The opening of vision*. New York: Routledge.
MacDonald, J. (1998). *The Arctic sky: Inuit astronomy, star lore and legend*. Toronto: Royal Ontario Museum.
Madison, G. B. (1990). *The hermeneutics of modernity*. Bloomington: Indiana.
Millar, R. (1998). Rhetoric and reality: What practical work in science education is *really* for. In J. Wellington (Ed.), *Practical work in school science: Which way now?* (pp. 16–31). New York: Routledge.
Norris, C. (1990). *What's wrong with postmodernism?* Baltimore: Johns Hopkins.
Nye, B. (2000). *About the show*. Available: http://www.nyelabs.com/open NyeLabs.html
Sagan, C. (1995). *The demon-haunted world*. New York: Random House.
Santayana, G. (1983). *Reason in science*. New York: Dover. Original work published 1906.
Science Council of Canada. (1984). *Summary of Report 36: Science for every student*. Hull, Québec: Canadian Government Publishing Centre.

Staal, J. D. W. (1988). *The new patterns in the sky: Myths and legends of the stars*. Blackwood, VA: McDonald & Woodward.

Yager, R. E., & Penick, J. (1986). What students say about science teaching and science teachers. *Science Education*, 68 (2), 143–152.

Yager, R. E., & Yager, S. (1985). Changes in perceptions of science for third, seventh, and eleventh grade students. *Science Education*, 22 (4), 347–358.

Zeidler, D. L. (1984). Moral issues and social policy in science education: Closing the literacy gap. *Science Education* 68 (4), 411–419.

Chapter Four
Serres Bugs the Curriculum
Marla Morris

My life as a classical pianist ended when I was in my early twenties. Tendonitis ruined my right arm. At thirty-eight I still cannot play a piano, but I can play a synthesizer. The weight of the keys is much more manageable than a piano, but my arm hurts if I play too much. The body has a memory. A musician spends a lot of time thinking about sound and its counterpart, silence. Rests, in music, help shape a round sound, give breath and grace to a phrase. The grace that silence brings is mystical. Silence seems to be that space that coincides with a sense of profound calm, an unutterable nothingness. Merleau-Ponty (1968), in *The Visible and the Invisible*, suggests that silence "coincides" with our very being. "The philosopher speaks, but this is a weakness...he should keep silent, coincide in silence, and rejoin in Being..." (p.125). Is silence Being or Nothingness? Is it both? Is it neither? Maurice Blanchot (1987) comments that "the more silence there is, the more it changes into a clamor. Silence, silence makes so much noise, perpetual agitation of the calm..." (p.86). Blanchot may be right here, for perhaps what I listen to in-between the spaces of sound is not silence after all but noise.

Noise brings me to my encounter with the writings of Michel Serres. Serres might suggest that silence does not exist at all. All there is, is noise. Being, Nothingness, Time, Space, Music = noise. My reading of Michel Serres has profoundly shattered my thinking about silence. For me, Serres' work has been a revelation of sorts. I no longer think of rests in music, for instance, as silence. I have grown more attentive to the constant background noise of my cat crunching, a lawnmower mowing, my thoughts grumbling. Grumble, rumble. Even in the dead of night, silence grows noisy. Ringing in my ears is what I hear, the air-conditioner kicking on, my cat knocking things off of my dresser. That silent nothingness I coveted for so long was an illusion, a trick of the tale. Like a shape-shifter, noisy shadows fill the empty spaces in the stairwell, on the wall, under the floorboards.

I come to Michel Serres' texts not as a scientist, nor as a philosopher. I read Serres' texts through the ears of a curriculum theorist. What I hear in Serres' texts is interference (noise) around what it might mean to be an educated person. Curriculum is not about neat and tidy interrelations between students, teachers, and texts. Rather, curriculum might become noisy. Lived experience in schools and universities might be considered interference.

Tracing Serres' work through *Hermes* (1982a), *Parasite* (1982b), *The Natural Contract* (1996), *The Troubadour of Knowledge* (1997), *Michel Serres with Bruno Latour: Conversations on Science and Culture* (1998a), and *Genesis* (1998b), I will explore what I call Serres' onto-theology of noise as the genesis of being (ontology), and as the ineffable, primordial chaos of the past and the now (theology). Further, I will discuss Serres' embodied noises in bugs, beings, and ghosts who utter a relational and fragile epistemology. Finally, I will examine what happens if noises and bugs are squashed. Serres offers to us a world of return to the sacredness of an onto-theology of noises and bugs. Serres bugs the curriculum (as an interference among teachers, students, and texts). And it is to this interference we should turn.

> In Serres' (1998a) *Conversations with Bruno Latour* he says, "I became highly sensitized to...the importance of Brillouin's work, of information theory in physics, and much later, of questions of turbulence...disorder and chaos...." (pp.11–12)

In order, then, to understand Serres' writings, it is first necessary to grapple generally with some of the concepts of information theory, or what Claude Shannon and Warren Weaver (1971) term the mathematical theory of communication. Scattered throughout Serres' writings, echoes of information theory can be heard.

Generally speaking, Claude Shannon's (1971) contribution to *The Mathematical Theory of Communication* seems to suggest that when making a telephone call, noise and static cannot and should not be eliminated from the line or channel. In a closed system, whereby a sender, a channel and a receiver interact, noise and interference are necessary and even "beneficial" (Weaver, 1971, p.19) to the overall system of communication. Weaver (1971) says Shannon's paper is "bizarre" (p.27) because it seems to indicate that noise increases information. It would seem that noise decreases information, but this is not the case. Moreover, when the message is transmitted over the channel, it changes, decays, and in effect becomes more uncertain due to interference and the breaking down of the initial signal. Thus, the more uncertain the message becomes, information actually increases. Therefore, noise and uncertainty increase information.

More specifically, Weaver (1971) says that information has a precise meaning for this theory. Information is "a measure of one's freedom of choice when one selects a message" (Weaver, 1971, pp.8–9). The more choices of information available to the sender the "greater is the uncertainty" (pp.8–9) that the receiver will hear the sender's intended message. Drawing on the concept of entropy, or the second law of thermodynamics, or what Weaver terms "shuffledness" (p.12), information gets dispersed and becomes "less and less organized" (p.12). William Paulson (1988) explains:

> The second law of thermodynamics, understood statistically and thereby translatable into a theory of information, states that in a closed system consisting of sender, channel, and receiver, information can lose, but not gain, specificity. (pp.66–67)

By the time the receiver hears the sender's message, it may not be exactly what the sender intended. Now, Paulson points out that noise, strictly speaking, is something that is added to the initial message that was not there to begin with.

> Noise may thus be the interruption of a signal, the pure and simple suppression of elements of a message, or it may be the introduction of elements of an extraneous message....(Paulson, 1988, p.67)

But even if noise seems to be added or "extraneous" it must be considered integral to the system. Paulson stresses that

> When we speak of such a system as closed, then, we do not exclude noise as something exterior; we include whatever noise is actually present in the channel....To exclude noise from a closed communication system would be equivalent to excluding molecular movement from a closed system as defined in thermodynamics; it would be equivalent to excluding time. (1988, pp.67–68)

Although information theory, according to Shannon and Weaver (1971), has nothing to do with "meaning," one could draw out many meaningful implications. My reading of information theory, ironically, echoes similar struggles in the field of hermeneutics, which deals with meaning. Friedrich Schleiermacher, who is associated with the school of romantic hermeneutics, taught that in order to do interpretive work, one had to get behind a text to reveal the hidden meaning and real intention of the author. Understanding a text, then, meant understanding what it was that the author intended to say. But over time those who practiced the art of hermeneutics became plagued with doubt. What was doubted was that interpreters could ever get behind a text to the intended meaning of the author. How can we know what someone intends? With the arrival of Hans-Georg Gadamer and Paul Ricoeur in the 1960s and 1970s, the art of hermeneutics began to change radically. Paul Ricoeur (1995), for instance, suggests that it is impossible to ever know the intended meaning of an author. Distance from a text in time and place creates uncertainty as to the author's meaning. And paradoxically, Ricoeur stresses that even if a reader is in close proximity in time and place to an author, uncertainty becomes more pronounced. When one is too close to an author, her writings, in fact, might seem even more obscure. Knowing an author clarifies little. Information theory would suggest that one of the reasons we cannot get behind a text or understand the intention of the author is that there is a lot of noise between what the author intends to say, what she actually says, and how the reader translates the text. The translation and interpretation are also noisy since the reader brings her psychological baggage to the text. Further, readers read texts in a noisy fashion because one text interrupts another. Texts are not read in isolation. And it is the noise of doing interpretive work that allows for generative understandings. Even though information theory is not hermeneutics, it is not incompatible with the ways in which we understand and read texts. And making a phone call could be considered a text. When I call someone I think my message is being understood. But who knows? The receiver on the other end may have the TV blaring, or the interference in the line may change the context of my chatter. Even in face-to-face discussion, it is hard to know what people understand in a discussion. When I teach I am often shocked to find out that my students haven't understood a word I've said. Or their interpretations are just plain bizarre.

Information theory is also not incompatible with what is called the theory of self-organizing systems. William Paulson (1988) explains:

> ...the question of noise and information in self-organizing systems...attempts to give a nontheological answer to the old question, why there is something rather than nothing? [this question was posed by Leibniz]. (p. 72)

In other words, the theory of self-organizing systems concerns how organisms come into being, come into existence, and draws on such concepts as chaos and order. Why is there something rather than nothing? It is not because of God, but because of the ways in which organisms grow in chaotic and ordered systems. Paulson (1988) explains that there are two kinds of "organized states" (p.72). One is called "ordered" (p.72) and has a "high degree of redundancy" [but] "a low level of information" (p.72). The other is called "complex" (p.72) and has a "low degree of redundancy" [and] "a high level of information" (p.72). Therefore, self-organizing systems "contain" (p.73) both redundancy and complexity (information), both order and chaos.

Chaos, order, redundancy, noise, interference, and information are concepts that creep into Michel Serres'writings in what I would term a "mystical poesis." He draws on scientific concepts but couches them in poetic language. He writes as a philosopher-poet. This is an important point to stress because the poetic noise of his texts, which for many scientists may seem odd or even inappropriate, fosters interdisciplinarity (poetics, science, and social criticism). This is what makes Serres accessible to the nonscientist and so unlikable by the conservative scientist who might be wedded to positivism. (Later in this paper I will address Serres' insistence on interdisciplinarity.) Another aspect of Serres' texts that probably maddens positivistic scientists is the mystical element that pervades his writings. Conservative scientists might not like this because mysticism is often associated with irrationalism. And irrationalism is plain anathema for scientists. But Serres has little patience for the fallout of rationalism, the fallout toward domination and destruction of the planet. Serres (1982b) comments "Yes, the divine is there; I touch it, these things are improbable miracles..." (p.46).

The noise, or combination of science, poesis and mysticism, is not, however, new. Serres echoes many of the pre-Socratic philosophers like Thales, Anaximander, Anaximenes, Pythagoras, and Heraclitus. These pre-Socratic philosophers of science ask questions about the world in poetic language tinged with mysticism. For instance, Thales thought that the world was ultimately made of water; water, in other words, was the fundamental principle of being. He also taught that magnets had souls because they could move things by attraction. Here is the mix of science and mysticism. Aristotle comments that

> Thales, too, seems, from what they relate, to have supposed that the soul was something inetic...the [magnesium] stone possesses soul because it moves iron. (cited in Kaufmann,1968, p.7)

Like Thales, Anaximander also believed that there was a principal element out of which all being came into existence. Anaximander said that the "apeiron" (cited in Kaufmann, 1968, p.8) causes things to be. This airy stuff also created "all the heavens" (Simplicius, cited in Kaufmann, p.8). Again we see the mix of science and mysticism.

Pythagoras, who invented the Pythagorean theorem, also started a religious sect. Members followed at least thirty-nine rules (Kaufmann, 1968). One of these rules, interestingly, was "Do not look in a mirror beside a lamp..." (cited in Kaufmann,1968, p.12).

Heraclitus, who believed the world was made of fire, flux, and chaos (perhaps the first chaos theorist) also thought that "The Lord whose oracle is in Delphi neither speaks nor concedes, but gives a sign" (cited in Kaufmann, 1968, p.17).

Plato kicked all poets out of his utopian Republic because he thought they were dangerous; he thought poets would corrupt young minds. Poets are not rational; they move people in an emotional way. Emotion, for Plato, is considered a dangerous thing. Emotion is tied up with intuition and appearance. Rationality, on the other hand, is the only way to get at the real. The only way to get above the divided line (the lower realm is emotion, the upper one is rationality), and the only way to get out of the cave (the cave symbolizes appearance and ignorance, outside the cave is light and the truth) is by way of rationality. It is with Plato that the West began to denigrate poets. I do think that scientists, who are still wedded to positivism, have inherited Plato's attitude. They don't like poets and they don't like anything that smacks of the irrational. As John Weaver, in the introduction to this book points out, scientists wedded to positivism call for language that is plain, simple, clear, precise, and unambiguous. But even when language seems to be plain, simple, clear, and unambiguous, it is still ambiguous. When one studies mathematical logic it becomes obvious that even mathematical symbols cannot capture language in a precise way. The more precise one attempts to be the more obscure the proposition becomes. A tautology ($A = A$) might be precise, but as Leibniz (1940) pointed out long ago, tautologies are trivial. That is, a phrase that does not generate anything new is basically meaningless. For example, if I say that my sister is my sibling, this is true. But this statement means nothing. It is already understood that a sister is a sibling. Leibniz would say that this statement is trivially true. And this may be as precise as language gets, but it means little. As soon as the predicate differs from the subject, language gets fuzzy. The call for unambiguous language by conversative scientists is naive. Little language is unambiguous. And if it seems precise its meaning is trivial.

I see Serres following the pre-Socratic poet-philosophers. Like these pre-Socratics, Serres mixes poetics, science, and mysticism. Unlike the pre-Socratics, Serres believes that the genesis of being is noise, not air

Serres Bugs the Curriculum

(the apeiron), water, or fire. Why is there something rather than nothing? Serres might say because of noise.

At this juncture I will take you, reader, on a journey through some of Michel Serres' noisy texts beginning with the beginning or *Genesis* (1998b). I argue that Serres offers to us an onto-theology of noise. Noise, for Serres, is the root of being (ontology) and the mystery of mysteries, the holy of holies. Noise is ultimately beyond conceptualization and enters into the theology of the mystical.

> Background noise may well be the ground of our being: It may be that our being is not at rest, it may be that it is not in motion, it may be that our being is disturbed. (Serres, 1998b, p.13)

The pre-Socratic philosopher Xenophanes argued that being is at rest. "He [being] always remains in the same place, not moving at all..."(cited in Kaufmann,1968, p.13). But Zeno argued that being is in motion, even though we cannot capture this motion in words. Serres, you see, counters both Xenophanes and Zeno by suggesting that being is "disturbed." Noise is not something external to us; rather it moves through us. Noise breaks through the inside/outside appearance of lived experience. Noise cuts through observer and that which she observes. Noise connects both observer and observed in complex ways; it undercuts the Cartesian and Sartrean split between the viewer and that which is viewed, the listener and that which is heard.

This is disturbing because we cannot, with any precision, say where we stand ontologically; we are mixed up in it all. There is no out there or in here. It is all mixed up in a noisy way. Scientists who are wedded to positivism might be uncomfortable with this ambiguous way of looking at lived experience because one can no longer say that what is out there is out there in-itself. That is, the in-itself is shattered. I am in the in-itself and it is in me. Serres, like Merleau-Ponty (1968), seems to suggest that bodies are mixed up in-the-world in complex ways, and so it becomes difficult to say what we see. Words are not mirrors to the world out there. And this is because the out there is in here. Lots of noise interferes with modernist notions of an ontology that is couched in dualistic language. I am interconnected with the world in which I live. I do not stand separate and apart from it.

Being is noise and it is disturbance, Serres tells us. Every time I walk into a classroom I create a disturbance; I bring with me a wave of turbulence, a jumble. Not only do I create a disturbance, I am one. Every time a new student enters the room, therefore, she creates a disturbance because she is disturbed already. This is why educating people is so difficult. A room full of 25 disturbances creates a very noisy situation.

Another disturbing thought is that in the beginning, that is, when the big bang happened, Serres suggests that it was not a bang at all. Serres (1998b) says, rather, "I am assuming that there was no big bang...there was and still is an inaccessible number of different noises" (p.61). It isn't that the primordial stuff blew up and wham, life began, but rather it is that life continually explodes. But many of these noises and explosions are "inaccessible." High–pitched sounds like the kind ultra violet rays put out bother some people more than others. I think it is no coincidence that teaching under these lights gives me headaches. Apparently, I can't stand the noise. But for other people these noises are not bothersome because they can't hear them in the first place.

Ultimately, for Serres (1998b), noise is beyond representation, beyond words. He comments that "We do not have a concept for it..." (p.67). It is that mysterious, primordial stuff that births us, drives us crazy, is everywhere the unnameable, ineffable life force like mana. Serres' onto-theology, or mystical writings on being, suggest that lived experience is never without noise because it is noise. And noise, says Serres (1998b), is "the multiple" (p.22), that is, it is not reducible to one thing. Noise is not a substance or a thing. It is fluid. It is

> ...raucous, anarchic, variegated, tiger-striped, zebra-streaked, jumbled-up, mixed up...criss-crossed by myriad colors and myriad shapes....(Serres, 1998b, p.22)

Noise is mixed up in zebras and tigers and in all kinds of colors. Serres stresses that we should not try to categorize noise. It is always much more complex than we might think. And so, here, Serres moves beyond information theory because he does not want to define noise strictly. Recall, information theorists Shannon and Weaver (1971) suggest that noise must be strictly defined. But Serres does not do this. In fact he unleashes noise as zebras and tigers ride off into the midst; noise becomes that fuzzy something which is sacred. "In the beginning is the echo: murmur" (Serres, 1998b, p.70).

Not only is noise the echo and murmur, it is embodied. And these embodiments are relational. Noise is embodied in bugs, beings, and ghosts. Serres (1982b) stresses that these creatures are interconnected in a "system of relations" (p.8). It is in a book called *The Parasite* (1982b) where he works with ideas around interrelations and epistemology. Playfully, Serres says the city rat and the country rat scurry off when they hear a noise at the door. It was only a noise, but it was also a message, a bit of information producing panic: An interruption, a...parasite...who produces disorder and who generates a different order. (1982b, p.3)

City rats and country rats (beings) are interrupted by parasites (bugs) who creep in by making noise. So the parasite is the one who is an interrupter, no doubt a something that gets into one's system, breaking it down by living on it. And yet, the more it breaks it down, the more it builds up a different order. Parasites "generate a different order" (Serres, 1982b, p.3). Thus, the parasite who interrupts the city rat and country rat changes their relation.

When a new student walks into the classroom she creates a disturbance because she is one. She bugs the curriculum. Education is a process whereby we feed off of one another, we interrupt taken-for-granted knowings, we generate new orders out of disorders. Education is not a smooth, easy process. It is buggy and bugged by the complexities and ambiguities of lived experience and uncertainty. Education is about interruption. If I don't interrupt my students' ways of thinking, I've done nothing. Our knowings are affected, infected, effected by the numbers and kinds of disturbances, or parasites, that arrive on the scene. Serres calls the parasite "a new epistemology..." (1982b, p.36). The parasite "at the intersection of relations" (p.43), is all knowing, relational, and fragile, because every time a parasite walks in, a new disorder and order arrive on the scene. Beings, then, thrive on a high amount of interrelation in order to know at all. Parasitical knowing turns on the recognition of the disturbed and turbulent, the redundant and the ordered. These paradoxical knowings feedoff of each other in noisy ways. But the parasite is not the thing-in-itself. As Serres points out "the meaning of the prefix para- in the word parasite...is on the side, next to, shifted, it is not the thing, but on its relation" (1998b, p.38). And sometimes parasites "prevent it [a message] from being heard, sometimes from being sent" (p.11). Thus, in our interrelations there are times that understanding will not happen; there are moments when we will be stumped by the issue at hand, as a parasite walks in the room and throws us for a loop. But confounded thinking is part of the process of knowing and can even generate new kinds of knowings through frustration. It is okay to feel frustrated, I tell my students. I know that they are beginning a new journey when they feel discomfort. It is when they are comfortable that I worry.

There is an element of unknowing in the parasitical relations with others' knowings. And sometimes these other knowings are ghostly. "...the Paraclete, [is] the common name of the Holy Ghost, the third person. He intervenes, interrupts, comes in through walls....He is the gift..." (1982b, p.46). Ghostly knowings, mysterious knowings are slippery, moving through walls. Perhaps we become haunted by ghostly knowings. How do we come to know what we do after all? Does knowing come through some sort of ghostly absence/presence? When I am interrupted by a ghostly presence in my knowings, I may be pushed onto a different track, a different way of understanding life. From ghostly knowings appear intuitions, hunches, flashes of insight like those of Zen disciples. These ghostly insights cannot be explained by reason. Even Aristotle, the rationalist that he was, stressed that we must develop our sense of intuition if we are to understand anything. But intuition is ghostly. We cannot say exactly what it is. And it seems some people are more intuitive than others.

Music is ghostly too. It interrupts in ghostly ways. Melodies keep coming back, even when one doesn't invite them. Rickie Lee Jones sings a song called "Ghosty Head." I often listen to "Ghosty Head" on my ride home from LSU. Listening to this song, I am interrupted in my knowings, feelings. I think differently because of this song. Why this is I couldn't say: it's ghosty. What is a Ghosty Head anyway? Who knows? When one is opened to ghosty intuitions and hunches, interruptions brought about even by music and color, creative thinking is born. Creativity does not come from drilling and memorizing, repeating the words of others. Creativity is not unleashed when one is closed to mystery. Knowledge is a ghosty experience that is not only lived in the head but in the heart. And it is through frustration and ghosty experiences that knowings change. Knowledge is slippery, fragile, and interrelational. It is not cut off from others' knowings and everyday experience. Serres stresses that "Knowledge is never cut up into crystalline continents...but it is like a group of oceans" (1997, p.56).

But ghostly, buggy knowings can be crushed. "People laugh, the parasite is expelled, he is made fun of, he is beaten..." (1982b, p.35). And people laugh at the parasite when they are steeped in a Cartesian mindset that reduces the complex to the simple; when they exclude anything that sounds antirational, (ear)rational, or irrational. Cartesian rationality, for Serres, has produced little more than that quest for "mastery and possession" (1996, p.32). Descartes has won over the scientific world and has eventually seeped into all academic disciplines

Serres Bugs the Curriculum

in one way or another, by offering only a clear and distinct method which would arrive at certitude about the world. Descartes (1940) said, "if we reduce involved and obscure propositions step-by-step to simpler ones..." (p.28) then we would "attain complete scientific knowledge" (p.32). But is knowledge ever complete? Can it predict and control the uncertainties of lived experience? Descartes (1940) believed that "all science is certain and evident knowledge" (p.18). As against Descartes, Brillouin (1964), like Serres, says "Let us candidly admit that we know nothing with certainty..." (viii). Descartes (1940) says that knowledge is "continuous—nowhere interrupted" (p.32).

Knowledge that would not be interrupted is linear, progressive, beautiful, harmonious, and whole. Serres' entire project, as against Descartes, is to interrupt the illusion of a linear, progressive, beautiful, and harmonious knowledge. Knowledge is not gotten at by thinking in isolation either. Descartes' entire epistemology is based around the individual who meditates in front of the fireplace and doubts everything until certainty about self, world, and God becomes clear and distinct. Knowledge for Descartes is gotten by the self and for the self. And there is nothing uncertain about it. Serres suggests that Cartesian "arrogance" (1998b, p.5) remains in the scientific community. It is arrogant to think one can know with certainty, predict and control, master and tame the wild world in which we live. Cartesian arrogance squashes anything that appears uncertain and ambiguous. And Serres says that this arrogance springs ultimately from fear. Serres comments that, "Fear comes from the swarming, the tick, the dread multiplies like flies....Rationality was born of this terror" (1998b, pp.66–67).

The residues of Cartesian logic creep into the scientific community in many ways. One form turns on "surveillance" and "agreement" (1996, p.45). Serres suggests that, at the end of the day, scientific communities are guided by a sort of social contract that allows scientists to engage only in a very particular discourse which produces "obedient specialists or ignoramuses..." (1996, p.46). And scientists may become narrow and ignorant because, as Serres points out, they follow a sort of Cartesian rationality that only allows for certain ways of thinking and talking, while squashing anything that seems to bug reason or go against the grain of traditional ways of doing science. Moreover, this narrow appeal to rationalism has created, in part, a dangerous world for all of us. Serres remarks that

> The successive rise of the sciences and their affiliated technologies, each of which, at the apex of its power, comes close to mortal danger—atom and bomb,

chemistry and environment, genetics and bioethics—these crises bring back the demand for prudence. (Serres, 1996, p.93)

Aristotle (1984), in the *Nichomachean Ethics*, argued that all virtues spring from prudence. But prudence is not the same as rationality. In fact, rationality, at times, opposes prudence. Rationality is quick and calculating; prudence is deliberative and slow. Prudence can lead to wisdom and can keep reason from going astray. Aristotle suggests that it is difficult to develop prudence. It takes time to take time, to be deliberative, to worry. Prudence, I think, is born of worry. And Serres says that what worries him is that we are running out of time. Scientists, who work without any sense of prudence or ethics, are on a path toward blowing up the world. Science "enters a dynamic of madness" (1997, p.122) when it thinks it is the only right and true path, when it remains cut off from the ethical implications of its work.

Serres urges scientists to re-think and re-consider the ways in which they go about working and understanding. He calls for a parasitical wisdom that draws on both literature and the sciences. Science cannot avoid its complicity with violence if it remains trapped behind the fences of Cartesian rationality and specialization. The more disciplines isolate themselves from each other, the more dangerous they become. Serres calls for "cross-breeding" (1998a, p.27). When the parasite is squashed, when the oddities of knowing and being are not heard, "the chased entity...always returns" (1982b, p.78). Ultimately, Serres suggests that university systems breed the rational disease of separation, classification, specialization, and hierarchy which have created arrogance and madness. However, Serres is not the first to call for a move toward interdisciplinarity.

William Paulson (1988) reminds us that C.P. Snow warned against what he called "the two cultures" (cited in Paulson p.4). Paulson tells us that "Snow denounced the consequences for intellectual ignorance of those who live by literature and those who live by science" (1988, p.4). The difficulty, of course, for many academics, is that within their own discipline there is already too much to know and understand. One might ask how a scientist, for instance, could possibly have the time to read literature? Or how could an English professor be expected to understand chaos theory? The academy demands a high number of publications at research institutions, which forces professors to stick to their own turf. Publications outside the major field of research do not count for many. So the academy has created a problem of specialization. And specialization does create a certain kind of scholar. Although interdisciplinary studies are popping up (cultural studies, religious studies, women's studies), still the academy, for the most part, seems to value specialization and relegates "studies" disciplines to the lower ranks of respect. As John Weaver points out in the introduction to this book, there is a tremendous backlash against the cultural studies of science movement because many conservatives feel that it is ruining 'pure,' 'real' science, since it seems irrational and silly. But Serres teaches, that is in spite of the difficulties one might encounter in doing the work of interdisciplinarity, in spite of the conservative backlash,

> one can no longer be [considered] educated without the hard sciences, without the history of science, of technology....without law or philosophy, without the history of religions or literatures. (1997, p.71)

Scientists who refuse, then, to grapple with ideas outside of their specialization are not really educated after all. And humanities professors who do not grapple with the sciences are not educated either. Being educated requires a great deal of crossing-over into fuzzy terrain, into new and different regions of knowledge. It seems absurd to me that even many philosophers who are supposed to be generalists are afraid to do work outside of their area of specialization. John Dewey, I think, was the last philosopher who really dared take the leap; he wrote about everything. What is wrong with that? Are we educated people if our knowledges are so narrow that we cannot even talk across disciplines? Do we have to spin our exegesis around the head of a pin? Or can we cross galaxies and do wild-eyed interpretive work?

I consider myself to be one of those wild-eyed interpreters. I am lucky because my discipline is curriculum theory. And curriculum theorists are generalists who are encouraged, after Dewey, to take the leap and draw on many different fields of inquiry. Now, I'm not saying that doing curriculum theory is easy. On the contrary, it is extremely difficult because one is expected to have a very broad knowledge base. And I admit that sometimes I feel like Sartres' (1964) character in *Nausea*, who was called the self-taught man because upon arrival at the library he begins his research in the stacks with the letter 'A.' Today the self-taught man will read authors whose names begin with the letter 'A'; tomorrow he reads books written by authors whose names begin with the letter 'B.' So on and so forth ad nauseum, ad infinitum. It is frustrating to feel as if one has to know everything and anything as a scholar, but for me the sheer overwhelming feeling of being responsible for knowing a lot and reading a lot is exciting. Now, it is true that knowing a lot and reading a lot doesn't mean that one attains wisdom. Wisdom is something that cannot be gotten out of stockpiling knowledges. But it can't hurt to read everything one can. And this means reading texts that are outside of one's knowledge base. Yes, it is difficult to become what Serres (1997) calls a "troubadour of knowledge," but the willingness to cross over into many fields serves to trouble what it means to be erudite. Erudition, for Serres, means being broad, not narrow. And as a curriculum theorist, I think he is right on this. I think being broad makes for better scholarship.

Science educators might learn from curriculum theorists; I think we really do work after the fashion of Serres. We work like "weavers" (1982a, p. 52). Serres remarks that

> one must find the Weaver, the proto-worker of space, the prosopopeia of topology and nodes, the Weaver who works locally to join two worlds that are separated...rejoins the rational, the irrational, namely the speakable and the unspeakable...(1982a, p.52).

Weavers frustrate the old notion of 'expert' and specialization because they dare to traverse boundaries. Weavers are heretics who take risks. Institutions which squash and expel heretics and weavers suffer in violent ways from the purifying effects of de-bugging. We must continually bug the curriculum, pushing back boundaries. It is interference into old ways of thinking that allows us to return to that place which values the sacredness of an onto-theology of bugs and noise. Serres bugs the curriculum. Serres offers to us musical, poetic, noisy texts so that we might listen in new mystical keys. Serres is a weaver of noise. Science educators, curriculum theorists, artists, musicians, mathematicians, and philosophers might all become interrupters, noisemakers, bugging academe to change the sound of education.

References

Aristotle. (1968). From *De anima*. In W. Kaufmann (Ed.), *Philosophic classics: Thales to Ockham* (p.7). Englewood Cliffs, NJ: Prentice-Hall.

Aristotle. (1984). *Nichomachean ethics*. (H. G. Apostle, Trans.) Grinnell, IA: The Peripatetic.

Blanchot, M. (1987). *The last man*. (L. David, Trans.) New York: Columbia University.

Brillouin, L. (1964). *Scientific uncertainty and information*. New York: Academic.

Descartes, R. (1940). Selections from the rules for the direction of the understanding. In T.V. Smith & M. Grene (Eds.), *Philosophers speak for themselves: From Descartes to Locke* (pp.16–49). Chicago: The University of Chicago.

Heraclitus. (1968). No title. In W. Kaufmann (Ed.), *Philosophic classics: Thales to Ockham* (p.14-17). Englewood Cliffs, NJ: Prentice-Hall.

Leibniz, G. W. (1940). First truths. In T.V. Smith & M. Grene (Eds.), *Philosophers speak for themselves: From Descartes to Locke* (pp.300–306). Chicago: The University of Chicago.

Merleau-Ponty, M. (1968). *The visible and the invisible*. Evanston: Northwestern University.

Paulson, W. (1988). *The noise of culture: Literary texts in a world of information*. Ithaca: Cornell University.

Pythagoras. (1968). No title. In W. Kaufmann (Ed.), *Philosophic classics: Thales to Ockham* (pp.10-12). Englewood Cliffs, NJ: Prentice-Hall.

Ricoeur, P. (1995). *Hermeneutics and the human sciences* (J.B. Thompson, Trans.). New York: Cambridge University.

Sartre, J. P. (1964). *Nausea*. New York: A New Directions Paperbook.

Serres, M. (1982a) *Hermes: Literature, science, philosophy*. (J.V. Havari & D. F. Bell, Trans.) Baltimore: The Johns Hopkins University.

Serres, M. (1982b). *The parasite*. (L. R. Schehr, Trans.) Baltimore: The Johns Hopkins University.

Serres, M. (1996). *The natural contract*. (E. MacArthur & W. Paulson, Trans.) Ann Arbor: The University of Michigan Press.

Serres, M. (1997). *The troubadour of knowledge*. (S. Favia Glaser & W. Paulson, Trans.) Ann Arbor: The University of Michigan.

Serres, M. (1998a). *Michel Serres with Bruno Latour: Conversations on science and culture*. (R. Lapidus, Trans.) Ann Arbor: The University of Michigan.

Serres, M. (1998b). *Genesis*. (G. James & J. Nielson, Trans.) Ann Arbor: The University of Michigan.

Shannon, C. E., & Weaver, W. (1971). *The mathematical theory of communication*. Chicago: The University of Illinois.

Simplicus. (1968). From *Physis*. In W. Kaufmann (Ed.), *Philosophic classics: Thales to Ockham* (p.8). Englewood Cliffs, NJ: Prentice-Hall.

Weaver, J. (2001). *(Post) modern science (education):Propositions, and alternative paths*. New York: Peter Lang.

Weaver, W. (1971). Recent contributions to the mathematical theory of communication. In C. E. Shannon & W. Weaver, *The mathematical theory of communication*. (pp. 1–28) Chicago: The University of Illinois.

Xenophanes. (1968). No title. In W. Kaufmann (Ed.), *Philosophic classics: Thales to Ockham* (p. 13). Englewood Cliffs, NJ: Prentice-Hall.

Zeno. (1968). No title. In W. Kaufmann (Ed.), *Philosophic classics: Thales to Ockham* (pp.22–31). Englewood Cliffs, NJ: Prentice-Hall.

Part Two
Pastiche Science: Bringing Cultural Studies of Science to Education and Education to the Cultural Studies of Science
Peter Appelbaum

Pedagogies of Sciences

A provocative debate about the nature and context of science has emerged within the last decade, throwing "science" into a Sargasso Sea of social and cultural context (Gross & Levitt; Gross, Levitt, & Lewis; Haraway; Holton; Latour; Ross; Serres; Sokal & Bricmont). The clamor effectively articulates important controversies and points of conflict about the approach to, and nature of, science and scientific efforts. However, participants do not talk about the pedagogy of science; players in these debates undertheorize education and inadequately address educational institutions of science. Meanwhile, those participating in or studying the so-called "science wars" are curiously absent from contemporary educational studies. And educators often accept a stereotyped and monolithic perspective on science. This section of the book seeks to respond to both of these empty spaces. We seek to pull science into current discussions of postmodern pedagogy and policy, while at the same time placing educational studies at the center of debate over controversies in science and postmodern science. We take as our starting points cultural studies of science, critics of these critiques of science, and contemporary theoretical orientations to educational studies. But we go further and describe what they have been doing now, as an entry into the ongoing creation of a postmodern education of science. We include both mathematics and science in our analyses because the general debates over science studies include both mathematics and science, and because mathematics is so often lumped in with science as "the language of science" or the "filter" through which potential science flows.

When critics of the cultural studies of science do address education, they express anxiety about a loss of traditional science in efforts to dilute science content. A good example is *A House Built on Sand: Exposing Postmodernist Myths About Science,* edited by Norella

Koertge (1998). This collection of key scholars in current debates about the meaning of cultural studies of science is pretty much a one-sided trashing of the cultural studies of science by traditionalists in science. In her own chapter, "Postmodernisms and the Problem of Scientific Literacy," Koertge characterizes "postmodernist accounts of science" as prescribing particular transformations of science education and a fundamental redefinition of scientific literacy. She declares that works such as Collins and Pinch's *The Golem* (1993), Sue Rosser's *Female-Friendly Science* (1990) and *Teaching the Majority* (1995), and Andrew Ross' *Strange Weather* (1994) have a common project of revolutionizing science pedagogy. Her argument seems to go something like this:

> I assume science practice as done by the popular image of scientist that I hold is good. I then believe that people could not understand what I think of as science unless they study traditional science. Now, there are people who describe what science is in an anthropological way, or through an analysis of the discourse and rhetoric of science, and these people describe some pretty horrific things. And there are also a few people at the college level offering courses in this stuff! What would happen if these courses became thought of as the "real" study of science? This indeed is an even more horrific nightmare! So I will ridicule what these people say and do as never being able to be called science according to my own definition, and then you too will call all of those involved in the cultural studies of science "Fools."

The fact that these authors do not claim to be scientists but rather observers of science is bypassed in an effort to "save" science.

But how does this discussion turn into a debate about science *education*? A leap from science to cultural studies of science, all of which might be "taught" in various departments at a university, becomes a transformation of science as we know it into postmodern science *education*. What do we hear from the people in education on this matter? Mostly silence. Perhaps it is because the cultural studies of science have occurred at the university level, where educational studies are denigrated to a nondiscipline on most campuses. Whatever contributions educationists might make have gone unheard. Yet it seems that there has been very little effort on the part of educators to grapple with the implications of the cultural studies of science. Perhaps it is because most of the people in educational studies have a social foundation (read *social* foundations background) that they are not comfortable talking about science. In general, science pedagogy has been left to the science education people who teach methods, and there has been only occasional analysis within education fields of science as a cultural phenomenon.

Pastiche Science

Also, teachers of science do not get exposure to the cultural studies of science and the implications for the teaching and learning of science, since this sort of critique of science is outside of the departments of science, and indeed is not the same thing as the science content they learn in their science courses. Science educators are mostly informed by traditional courses and perspectives in science. Discussions of curriculum end up being mostly about science content, the proper sequence of content, a perceived hierarchy of concepts and skills, and factual knowledge that becomes a foundation for later study. Only rarely do we talk about controversies about the nature of science in education courses that deal with the teaching of science. Collins and Pinch, Rosser, and Ross are not commonly read as "science" in science content courses. What a sad state for educational studies! Yet what an opportunity for us now to enter the general discussion of science education *as* people who do educational studies and respond to the perspective illustrated by Koertge. There is a modernist pedagogy of a modernist science, illustrated by what comes most immediately to mind when we picture a classroom in which science is being taught or learned. Here we expand our notions of science pedagogy to include postmodern pedagogies of modernist science, modernist pedagogies of postmodern science, and postmodern pedagogies of a postmodern science.

It is the case that mathematics and science educators have taken on some of the rhetoric of science studies. For example, there is attention to the importance of relationships among science, technology, society, and human values, and the importance of discourse and 'communities of inquiry' in the classroom. Assessment has expanded to include students' attitudes and relationships with the disciplines or topics. Yet these examples are representational in being so steeped in the rhetoric of standards, workplace readiness, and skill attainment that most of the flavor of cultural studies has been lost. Nevertheless, critics of science studies are very nervous when they see the words in print or hear about educators who focus on culturally constructed knowledges of gender, objectivity, social context, or relevance to students' lives and popular culture. Cultural theorists recognize these things as "signifiers"; the critics have fears, including a loss of privilege and cultural capital.

Koertge (1996) earlier wrote of feminist epistemologies and the pedagogies they support as thoroughly undermining science-as-we-know-it. This essay, "Feminist Epistemology: Stalking an Un-Dead Horse," again demonstrates a common set of fears and defenses of the status that scientists hold after going through their apprenticeship into traditional science practices. The presumption is that a postmodern pedagogy would always create a postmodern science that defies some core norms. We must ask, first, whether or not a postmodern *education* could indeed be most appropriate for achieving a *traditional* form of science knowledge and practices—perhaps more effectively than other, common sense pedagogies. And second, we need to reflect on the potential impacts of a postmodern or other pedagogy that promotes the creation of a truly radical version of science itself. Koertge clearly worries about the second scenario: she believes that feminist and other epistemologies are so incompatible with the science that she has learned to embrace that they will destroy the possibility of continuing such a science; like other critics of the cultural studies of science, she sees the cultural studies of science as a haunting specter of the death of science. Surely a postmodern science education would not kill science. But we might indeed find a different science that Koertge would not name science. A postmodern science education must examine the norms of science—the rules by which traditional science stakes its claim to have a story to tell—and that examination must place into question which people have the most to gain by the variety of challenges to, or efforts to save, these norms. Students do not necessarily act like adult scientists in order to learn science. On the other hand, what we think of as a scientist that should be modeled by a student "acting like a scientist" is not without its own complications (Gough, 1998). And, indeed, I rarely find students in science classes enhancing their propensity to interpret the world as a scientist. Instead, they spend most of their time learning to parrot already-developed techniques and applications of science. Furthermore, science education as it is currently practiced is not necessarily about being an apprentice scientist but instead promotes a variety of subject positions with relation to the content of science, including critical citizenship (decisions that use scientific information), responding to being the object of scientific study, and the role of science in social business policy (Weinstein, 1996).

I bring up Koertge's complaints about feminist epistemology because I believe what she highlights from this epistemology help mathematics and science educators identify just what pedagogical strategies would most serve their needs as educators of science, whether or not they want to transform or preserve the science itself. We can use what she criticizes as the initiation of an effective postmodern education. Koertge (1996) sees feminist pedagogy as a direct assault on the ideas of Talcott Parsons and Robert Merton. "When we turn to radical feminist critiques of science based on feminist epistemology, we find a repudiation of the ideals themselves" (Koertge, 1996, p.417). She uses the example of one certification program for science educators that requires a reading of Belenky et al.'s (1986) *Women's Ways of Knowing*, mischaracterizing this work as teaching the need to change science to meet the essential inadequacies of female minds. In her reading, such an approach necessitates a diminishment of the quality of science as science, thus underserving females in the long run. Of course, this is not at all what Belenky et al. want to say. They instead want to add to the narrow epistemologies (of science) in our repertoire to include richer comprehension of the possibilities for modes of knowing. The contemporary debate about postmodernism is so often framed in all or nothing terms (Kincheloe et al., 1999): we can either completely accept or completely reject Western modernism. This section of our book works to hold onto the value of modernist science while presenting images of currently existing practices that also enter into dialogue with postmodern perspectives on education and postmodern perspectives on science.

As we learn from John Dewey, beliefs are the key—they are that upon which we are prepared to act. But what we believe about science is not necessarily what we believe about how to learn science. Maybe this is a little too idealist, but the ideas drive action in this sense even (more likely) if we see blind adherence to practice as constructing beliefs implicitly; then change is driven in academics or intellectuals by an Enlightenment attitude that we need to rethink beliefs and then judge our practice based on these new beliefs. Viewing cognition as a process of knowledge production presages profound pedagogical changes. Postformal teachers (Kincheloe et al., 1999) facilitate interaction at the frontier where information of the science disciplines intersects with understandings and experiences that individuals carry with them to school; they help students to reinterpret their own lives and uncover new propensities as a result of their encounter with school. These teachers see their role as creators of situations where students' experiences can intersect with information gleaned from the academic disciplines. In contrast, if knowledge is viewed simply as an external body of information and codes of conduct and value, then the role of the teacher is to take this knowledge and insert it into the minds of students and to take these codes and police their students' behavior in accordance with the codes (Appelbaum, & Clark, in press). Evaluation procedures are intimately tied to the views we hold. Conceptual thinking would be discouraged in a modernist classroom, trivializing learning. Students would be evaluated at the lowest level of human thinking—the ability to memorize and mimic behavior. Unless students are moved to incorporate school information into their own lives, schooling will remain an unengaging rite of passage into adulthood, an experience of biding time until it is over (Appelbaum, 1999). Perhaps some examples of some potentially postmodern pedagogies of science could help us with both the beliefs and the practices.

What Is a Picture of (Post) Modern School (Science)?
Example One
We are in a first grade classroom. The class reads *Desert Giant* (Bash, 1989). Whole-class discussion explores the multiple perspectives represented in the story; students write their own version of the story, Figure:

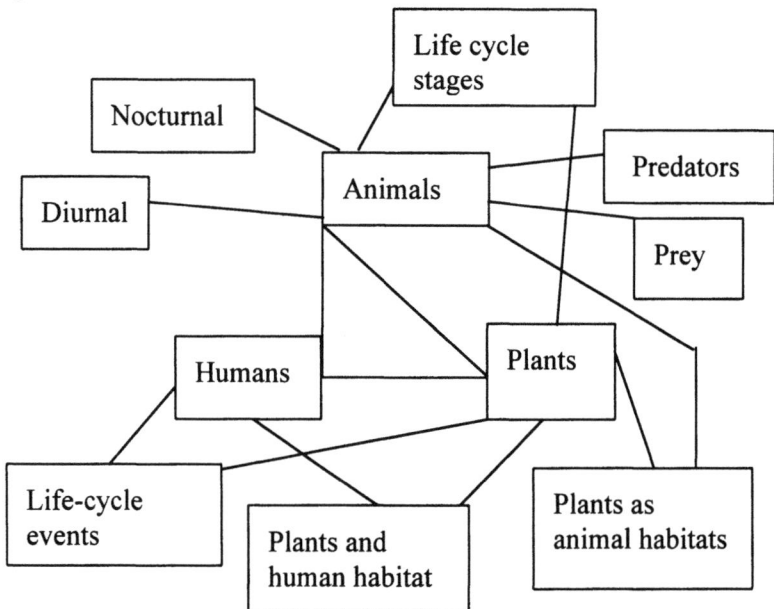

extracting one of these perspectives. Later in the week, students take on a perspective of their own choice for field study in which they record in their notebooks observations of an area near the school. They write stories from the perspective they chose. Then, in groups, they weave multiple-perspective narratives about the eco-system they observed.

> Objectivity redefined with values antithetical to the traditional meaning: observers should always remain emotionally connected to what they are studying; the richness of subjective experience should not be stripped away in the vain search for a lowest common denominator of objectivity; intuition should not play second fiddle to abstract, cold rationality/objectivity;

> knowledge is always perspectival and tied to local context, and the attempt to find an objective or 'God's Eye' point of view always ends up privileging the powerful. Then the playful curiosity so characteristic of so-called pure science must be replaced by an attitude of caring and commitment. (Koertge, p.417)

Koertge presents a view to scorn. Yet I believe that students learn best when they *care* about what they are learning, when what they do matters, when there is a purpose, a commitment to what they are doing. The navel-gazing elitist love of a playful detachment undermines internalization and connection to what is being studied. Students in this first-grade class are experiencing the roles of context and participation in the collection of observational data and field notes. And, while they may or may not be interested in ecosystems for their own sake, they are introduced to a way of making meaning out of an ecosystem that they are a part of. They are communicating their interpretation of this ecosystem to others who also have intimate relationships with the same ecosystem. A playful detachment might come about as students become fascinated with a particular aspect of science or as they become immersed in the process of an investigation, but this experience is serving the needs of a larger context of caring and engagement.

A fourth grade class is in the midst of a tree branch study. Over the course of a week they collect lists on large sheets of paper taped to the wall of the classroom: what they know and what they wonder. Groups form by interests growing out of the list of wonders and begin to collect data to help them guide an investigation about what they wonder about. Based on their initial data and consultation with other class members in whole-class meetings, the groups design research projects based on more refined data collection. All along, two students take on the job of anthropologists, and two others take on the job of the media. They study how the "scientists" are doing their work and give media reports, representing the work of the scientists to the class.

> Logic is a patriarchal device for brow-beating nonlinear thinking; since all knowledge is contextual, the search for generality is a form of imperialism; empirical validity must be tempered by moral and political appraisals. (Koertge, p.417)

Indeed, these fourth-graders are motivated to examine particulars of their own identification and rarely follow a linear process in developing their investigations. The representation of science in the usual textbook or hands-on activity rarely creates images of disorganization, illogic, intuition, and so on. In this classroom, the anthropologists and media reporters help students to reflect on the work of actual science. Now, rather than search for generality, students here might focus on particularities. They may create thought experiments, asking "what-if-not" questions about what they observe, moving into a realm of science fiction and fantasy. Problem posing might highlight the special method of 'What-if-not?', taking illogic as the road to local, contextual validity that sheds light on the general.

Example Two
Qishana reports to the class: "We needed to know, what do you have to add to 2/3 + 1/5 to get 1? (Walter, 1996) We found, after a lot of work, 2/3 + 1/5 + 2/15 =1…and then Shareefa laughed and said, we could've just multiplied the numerators and denominators instead of going through all that! Then we laughed and said, 'course not, that makes no sense. But Shareefa and I went off and looked for other combinations of fractions that we could solve the 'wrong way.' So we came up with…"Later, Roza asks, "So do we know now when multiplying works, and when it doesn't?" Shareefa: "We know it works when it works, and it doesn't work a lot; but your question—a good one—is still out there for studying more. We're thinking it's one of those questions that keeps us learning without ever finding the answer to the question."

Example Three
In Karen Grindall's fifth-grade classroom (Weaver & Grindall, 1998) students are using multimedia software to construct a story about the extinction of dinosaurs. They create two characters that join the dinosaurs and witness their extinction, one a mellow male hippie, the other a man who sings all the time.

Student 1: OK…the dinosaurs become extinct from a big explosion.
Student 2: No, they died from a cloud of dust.
Weaver: Could it have been both?
Student 2: No, I doubt it.
Weaver: Do we really know? Besides the issue is how will you create a story.
Student 1: Why don't we make our characters experience what we believe happened?
Weaver: Do you mean create two stories?
Student 1: Yes.
Weaver: Good idea, this way you leave it open as to how they actually died. You let the reader decide.
(Weaver & Grindall, 1998, p.244)

Pastiche Science

The hypermedia does not impose a modernist story structure of a beginning, middle, and end. The students were able to leave the story open-ended rather than try to reach a forced consensus in which only one possible storyline was acceptable; they were also able to morph identities, forming new identities that transcend modernist notions of time and space. Rather than stationed into fixed categories such as male or female, past or present, postmodern students move between them, constructing new markers that better capture complexity, ambiguity, and contradictions.

Example Four
Ms. Loella enters a unit on "ratio" with her class. She starts by saying, "O.K., we're going to start our unit on ratio, which is part of the 5th grade curriculum. I've laid out for you around the room different textbooks and activities that different companies and schools have used—they're all designed to teach you about ratio. My minimum expectation is that you explore several of the options and think about how they are similar and different. How do they differ in what they want you to think about? Later, I will ask you to develop a project of your own interest that can help you extend your knowledge. At that point, you'll be asked to represent what you've done and learned to three different audiences...." The students are used to these long introductions to a unit, and know that they will be asked to either make a presentation to an audience outside of their class or somehow communicate the importance of their activities to others. "...Your third-grade reading buddies, your parents and me, and a panel of three teachers that will help you judge the quality of your work."

> Community of a nonhierarchical sort is acceptable, but the rest of Merton's norms must go: a humane community would be based on trust, not skepticism; universalism should be replaced by standpoint theory, which says that reports are always to be understood as a product of the culture, gender, ethnicity, class of the observer who made them; no activity can be or should be disinterested. Quite the contrary, a commitment to correct political and social goals is to be encouraged. (Koertge, p.417)

Indeed the most vibrant pedagogical strategies are consistent with the "feminist" approach; the teacher and students *trust* each other and 'go with' a student's idea, regardless of logic, initial acceptance, or correctness. Karen Gallas (1994) describes a child that writes about liking rap music in his science journal—the teacher and student teacher do not see this as science. Gallas realizes she needs a new tack and asks him why rap is science. He says because it is "exciting." Throughout the year, he gets excited about all sorts of things and writes about them in his journal. By the later third of the year he is writing about 'exciting' animals and nature things and what he knows and conjectures about them, which looks like science to us. Bob Strachota (1996) takes a student's pseudo-conception or misconception and asks the class to take it as seriously as he does: if this is an explanation, if we begin with this theory, then what does it suggest to us, where does it take us? Ellie declares that an abalone is not a creature, because, "...if it was a creature, it would always be attacked because it would have to be like this (she held her hand with the open side up). It couldn't live the other way around because if a creature was in there it would fall out. Anything could easily kill it if it was like this (open side up)." Julius, reporting from a book, suggests the holes are where it goes to the bathroom. Ellie insists, "The holes must happen from the barnacles falling off." A series of ideas about the abalone shell are offered, all to be refuted by Ellie's consistent interpretation. Strachota reflects:

> ...misinformation rather than correct information stimulated the vigor and depth of the inquiry and, ultimately, led to a high level of understanding. Without Ellie's notion that the abalone wasn't a creature at all, we wouldn't have wondered how it could move with such a heavy shell, how it could stay in that shell, how it could stay safe, or what purpose those holes really served. Since I began working this way I have found that productive discussions often spring from misguided notions. (Strachota, 1996, p.52)

These classrooms are based on building a community of inquiry and demonstrate the importance of the social and cultural for all learning.

Example Five
Students in Pat Xave's class are told they have to use mathematics and science to learn how to juggle, or do magic tricks, or mix glazes for ceramics, or quilt (a) well enough to perform or display their work at the town library; and (b) well enough to help another person learn how. Students photograph themselves and their work at routine intervals of time with a digital camera, and using Kidpix Studio, create animations that represent their efforts over time.

The lesson for science *education* is that Parsons/Merton/Koertge is not to be applied until, in a *certain context*, a student develops a need for these norms...as a practicing scientist. When learning science, doing science, the feminist critique is apt, and may even lead to scientists aware of the critique, who are part of a cultural shift in the doing of science. But these people would not abandon science: they would still be doing (postmodern) science.

Example Six
Students are studying the work of Alexander Calder in art in fourth grade this year, and simple machines in science. So the art specialist teacher and the classroom teacher plan a team-teaching unit. Students form groups and use simple machines to create an "Energy Circus" exhibition inspired by Rube Goldberg contraptions. The exhibit is set up for a month in the lobby of the school, where every student in the school can participate on their way to something else.

Doing Science in School
How do we *do* (post) modern (science) education? The following chapters suggest important strategies. They take further the crucial features of a class room that I have begun to articulate:

- Science is not a distinct set of knowledge handed down, although its image of being so should be studied as a part of science education.
- Representation of the work of scientists and the usefulness of science is important in a postmodern classroom, including the representation of students as scientists.
- Illogic can be fruitful in learning science.
- The purpose is not necessarily to efficiently train students in obtaining answers to pre-given questions but more appropriately helps them develop ways to pose and critique questions.
- Audience is a feature of communication in a postmodern classroom.
- Science stories are not tied to modernist narratives with beginnings, middles, and consensus ends; we may know a lot about possibilities without drawing a conclusion.
- The pedagogical relationship between the student and the science is not best served by parroting a received view of science, but often requires 'unscientific' relationships of caring and connectedness.

In our first chapter, "Science Education through Situated Knowledge," Matthew Weinstein explores the possibilities of using the sociology and anthropology of science as a curricular basis for science education. He starts with the thought experiment of what a possible science education could speak to the needs of both the kin of the scientists and the kin of the subjects of the infamous Tuskegee Experiment. He draws on the idea of "situated knowledge" to talk about a multipositioned science education. A key feature of postmodern educational approaches to classroom practice, multipositionality can become a basic structure of lesson planning and classroom orchestration. And, as Weinstein notes, situated knowledge may therefore be a realization of a central goal in mainstream science education: a science for *all*.

Genre theory is another useful tool, as we search for ways to utilize the existing curricular materials available in our classrooms. Susan Gerofsky demonstrates the effectiveness of genre theory as she reflects on the teaching and learning of mathematics word problems. In doing so, she responds to a fundamental issue in the cultural studies of science, well articulated by Norman Levitt (1996). Levitt thinks "the problem" is really that some of us are afraid of the mathematics in both math and science. He thinks cultural studies of science and post-modern pedagogies of science and mathematics are really all about expressing resentment against mathematics and the reasoning and ideas it employs, because people can't understand it. In pop-psyche interpretations of postmodern critiques of 'science,' the wistful dream becomes, "if only more people, especially the 'well-educated'—had a different relationship with mathematics"—of course, upholding the privileging of "the" mathematics. Gerofsky understands both mathematics and genre theory, and asks us to construct word problems in school as parables or riddles rather than as disposable exercises (as they usually are). As parables worthy of longer and deeper contemplation, we might spend a week considering a single word problem in all its considerations and implications. As riddles, a playful and perhaps competitive spirit would be invoked; a pleasurable, oral culture approach to a recreational use of word problems would take the place of our present, very serious approach to evaluation of student written knowledge. As she writes, the simple suggestion that mathematical word problems be considered *as parables* or *as riddles*—the shift to the "as if" point of view that characterizes play and drama—may engender a shift in thinking and in educational practice.

Elaine Howes and Bill Rosenthal confront the issue of mathematics content and people's relationships with it. Lamenting the loss of "the infinite" in school science, school mathematics, and teacher education, they frame their desire to revive such study "postmodern." Modernism, they note, has made the infinite part of both mathematics and science a topic *non grata* in the face of the "empirical truth" that the infinite is the one (and only) mathematical topic with which just about everyone seems to be enthralled. Inherently paradoxical and contradictory, the infinite is the ultimate postmodern subject-object. They interrogate mathematics as a science of the infinite, mathematics as the language of science, the gendering of the infinite in mathematics-science, and the normalizing role of mathematics and science in "taming" the infinite via monotheism and patriarchy. The direction is not toward a conservative mathematics that saves or recreates science as postmodern but a postmodern conception of postmodern science.

In "Cookbook Classrooms; Cognitive Capitulation," Dave Pushkin points our attention to undergraduate education in science courses and its implication for pedagogy beyond these courses. What exactly could or should prospective teachers of science experience, whether in preparation for college–level teaching, secondary, middle, or elementary and early childhood? Pushkin articulates a concern for science education and prompts a response that is aware of our postmodern condition. Why are students losing interest in the teaching and learning of science? How might the pedagogy of science be undermining the potential for a vibrant science education? As Pushkin describes college science, students are forced to move "backward" from multiplistic thinking to a dualism that demands submission to a higher, unquestioned authority, and teachers are forced to limit access to knowledge in order to preserve students' apprenticeship.

As we consider the possibilities for a (post) modern (science) education, we need to examine the nature of supervision and administration of educational programs. Curriculum supervisors and principals need to understand the controversies and dilemmas in the teaching and learning of science, the role of culture in science pedagogy, and the relationships among popular culture, everyday life, and school science. Indeed, the very structures of supervision must support those aspects of classroom practice that are valuable. Jeffrey Glanz reminds us of the long history of supervision as striving to become a science itself and offers a metaphor that has led him to understand supervision in a postmodern society. He makes the case for supervision as *tofu*—as a permeable and adaptable substance that takes on the flavor of the foods that surround it. As tofu, unassuming yet nutritious, makes an ideal substitute for high–calorie foods, supervision as tofu also blends into the educational landscape to help provide needed services and assistance to teachers.

As refractor, I ask, "If current science education practices are simulacra (Baudrillard, 1990) of nineteenth century science, while cyberpunk art forms and popular culture more effectively communicate concepts of modern science (Gough, 1993), what does this mean for the science education of emerging teachers, for ongoing professional development, and for the pedagogies of educational studies? How might a conversation among scientists, colleagues who teach science methods, and those in curriculum theory unfold?" Treating both school science and educational studies as "theaters of representation" (Gough, 1998), these chapters could be refracted through the notion of "pastiche," at once a fragrant potpourri and cacophonous jumble of images, practices, and contexts of science and the pedagogies in and out of science. Common motifs include:

- concern for the cultural studies and anthropology of science
- interest in classrooms as theaters of representation
- analysis of curriculum as genre
- postmodern content in mathematics and science, and
- serious attention to the supervision and administration of (post) modern (science) education.

These chapters should be read in the context of two important directions for mathematics and science education: the rise of virtual reality and dynamic software environments, and the need for an educational practice that exhibits critical multiculturalism (Oakes & Lipton, 1999). In virtual reality environments, the goal is not to represent "reality" in a modernist sense, but to create a way of experiencing alternate realities. Commercial software packages such as SIM products or pinball construction programs, simulations of real-world phenomena, shift into an experimental mode that encourages explorations involving a change in the fundamental properties of a system—for example, the ecosystem of a natural habitat, or the gravity and elasticity of materials in a pinball game. In prevalent dynamic software environments, such as *Starlogo* and *Geometer's Sketchpad*, users fabricate a world of their own and then play around within it; concepts and models of worlds emerge in the context of exploration and invention. In both simulation and open environments, teachers and students have the opportunity to reflect on the metaphors that are the basis of their model of reality or new reality. Yet educational questions arise regarding the relationship that is constructed with the knowledge: are people who use simulations, metaphorical worlds, and new realities detached from the natural world, or are they more importantly able to critique how these relationships are constructed through the metaphors and analogies that they explore in the virtual reality environments?

In critical multiculturalism, I ask for critique of the narratives that are presented in school science. Any particular practice illustrated or fantasy posited in this and the following chapters should presume that the educators involved have asked, and will continue to ask, key questions:

- Who benefits from this version of the story?
- Whose prior knowledge and cultural experiences are best matched to the most important principles of the lesson? And whose are excluded?
- How will I get *students* to ask and answer these questions themselves?
- What (community) action projects will ask students to participate in local political processes?

As teachers and students begin to articulate these issues and examine their participation in the conflicts present within and around the cultural studies of science and science education, they may at times experience a modernist pedagogy of modernist science. And indeed I believe one should not study science without understanding its modernist "family of origins." However, they may experience at other times a postmodern pedagogy of science, or a modernist pedagogy of postmodern science, or, finally, a postmodern pedagogy of postmodern science. And in schools where the teachers and students are open to any of these, there will be in place the multiplicity of possibilities that is in itself a (post)modern (science) education. What does this mean for "science"? I believe it means that the pedagogy of science is as important, if not more so, than science education, that (post) modern (science) education needs to inform both the doing of science and the teaching and learning of science as much as the work of scientists has, in modernist education, informed science education. The postmodern "reality" of (post) modern (science) education is that it is in itself a field of emerging virtual realities; it is itself a pastiche of postmodern modernity, modern postmodernity, and postmodern postmodernity and requires of us the same refraction and critical multicultural attention as science studies have had the fortune to have lived through in the last couple of decades.

References

Appelbaum, P. (1995). *Popular culture, educational discourse, and mathematics.* Albany: State University of New York.

Appelbaum, P. (1999). Target: Number. In Kincheloe, J., et al. (eds.) *The Post-Formal Reader: Cognition and Education.* New York: Falmer, 423–448.

Appelbaum, P., & Clark, S. (in press). Science! Fun? A critical analysis of design/content/evaluation. *Journal of Curriculum Studies.*

Baudrillard, J. (1990). *Seduction.* New York: St. Martin's.

Bash, B. (1989). *Desert giant: The world of the saguaro cactus.* Boston: Little, Brown.

Belenky, M., Clinchy, F., McVicker, B., Goldberger, N. R., & Tarule, J. M. (1986). *Women's ways of knowing: The development of self, voice, and mind.* New York: Basic Books.

Collins, H. & Pinch, T. (1993). *The Golem: What everyone should know about science.* Cambridge: Cambridge University.

Gallas, K. (1994). *Languages of learning: How children talk, write, dance, draw, and sing their understanding of the world.* New York: Teachers College.

Gough, N. (1993). *Laboratories in fiction: science education and popular media.* Geelong, Victoria, Australia: Deakin University.

Gough, N. (1998). 'If This Were Played upon a Stage': School laboratory work as a theatre of representation. In J. Wellington (ed.), *Practical Work in School Science: Which Way Now?* London: Routledge.

Gross, P. & Levitt, N. (1997). *Higher superstition: The academic left and its quarrels with science.* Baltimore: Johns Hopkins.

Gross, P., Levitt, N., and Lewis, M. (Eds.) (1996). *The flight from science and reason.* New York: The New York Academy of Sciences.

Haraway, D. (1997). *Modest-witness@second-millennium.femaleman-meets-OncoMouse: Feminism and technoscience.* New York: Routledge.

Holton, G. (1996). *Einstein, history, and other passions: The rebellion against science at the end of the twentieth century.* New York: Addison-Wesley.

Kincheloe, J., Steinberg, S. & Hinchey, P. (eds.) (1999). *The post-formal reader: Cognition and education.* New York: Falmer.

Koertge, N. (1996). Feminist epistemology: stalking an un-dead horse. In Paul Gross and Norman Levitt, *The flight from science and reason.* Baltimore: Johns Hopkins University, 413–419.

Koertge, N. (Ed.) (1998). *A house built on sand: Exposing postmodernist myths about science.* New York: Oxford university.

Latour, B. (1996) *ARAMIS or the love of technology.* Cambridge: Harvard University.

Levitt, N. (1996). Mathematics as the step-child of culture. In Paul Gross and Norman Levitt, *The flight from science and reason.* Baltimore: Johns Hopkins University, 39–53.

Oakes, J. & Lipton, M. (1999). *Teaching to change the world.* New York: McGraw-Hill.

Ross, A. (1991). *Strange weather: Culture, science and technology in the age of limits.* New York: Verso.

Ross, A. (1994). *The Chicago gangster theory of life: Nature's debt to society.* New York: Verso.

Rosser, S. (1990). *Female-friendly science: Applying women's studies and theories to attract students.* New York: Teachers College.

Rosser, S. (1995). *Teaching the majority: Breaking the gender barrier in science, mathematics, & engineering.* New York: Teachers College.

Serres, M. & Latour, B. (1995). *Conversations on science, culture, and time.* Ann Arbor, MI: University of Michigan.

Sokal, A. & Bricmont, J. (1997). *Impostures intellectuelle.* Paris: Editions Odile.

Strachota, B. (1996). *On their side: Helping children take charge of their learning.* Greenfield, MA: Northeast Foundation for Children.

Walter, M. (1996). Curriculum topics through problem posing. In Rebecca Corwin et al. (eds) *Talking mathematics: Supporting children's voices.* Portsmouth, NH: Heinemann, 141–147.

Weaver, J. & Grindall, K. (1998). Surfing and getting wired in a fifth grade classroom: critical pedagogical methods and techno-culture. In Joe Kincheloe and Shirley Steinberg (eds.), *Unauthorized methods: Strategies for critical teaching.* New York: Routledge, 231–251.

Weinstein, M. (1996). Towards a cultural and critical science education. Paper presented at the annual meeting of the American Educational Research Association. New York, NY.

Chapter Five
Science Education Through Situated Knowledge
Matthew Weinstein

Introduction

The Tuskegee study did happen and other studies did happen. There's this aura, there's this feeling in the black community that these things happen to us...I believe their feelings are justified.—Dr. Isaac Powell, associate professor of urology, Wayne State University. Detroit News; 6/17/97

This paper is an exploration of some of the necessary conditions for a science education that would meet objectives of equity and social justice. It takes as its starting point and critical focus the NRC's National Science Education Standards (National Research Council, 1996, hereafter NSES or the *Standards*), which form a guide for so much of current policy in science education. On the one hand it examines how dreams of equity, as they appear in the *Standards*, are interrupted by what I call scientific monotheism, in short the belief that only scientists, their knowledge products, and an idealization of their operating procedures should represent science. It then goes on, however, and explores an alternative, what I title a multifocal approach to both science and science education. This derives from anthropological and cultural studies approaches to the sciences and argues that a plurality of knowledges and practices are present in most facets of the enterprises known as the sciences.

To make clear the limits of the *Standards* and the possibilities of a multifocal science education I will invoke a hypothetical classroom. This is a class located in Macon County, Alabama, and consists of children related to both the medical personnel who conducted the infamous Tuskegee experiment as well as children related to those who were experiments subjects. The Tuskegee study was carried out by the U.S.' Public Health Service in the 1930s through the early 1970s (when it was exposed) and involved 399 African American men who had late–stage syphilis as well as 201 control subjects. Its purpose was to observe the evolution of morbidity and mortality among the infected of the human subjects. No treatment was ever administered, though medical care had been promised to the subjects; no one was informed that they had syphilis nor were their partners notified (Jones, 1993).

> *Tuskegee happened because the scientists didn't value blacks, plain and simple.*—*Dr. Barbara Ross-Lee, Dean of Ohio University's College of Osteopathic Medicine,* Atlanta Journal and Constitution,; 8/16/97

The experiment has become in the 1990s a highly symbolized event, akin in its charge and emotional impact to the Holocaust. For instance, one of President Clinton's choices for surgeon general, Henry W. Foster Jr., was defeated in 1995 in part because of "unproven allegations that he knew about the Tuskegee experiment but failed to act" (Harris & Fletcher, 1997). In 1995, Clinton invited the five survivors of the experiment to Washington to receive a formal apology. In addition, he provided a $200,000 grant to Tuskegee University to help build a center on bioethics and research and directed Donna Shalala, Secretary of Health and Human Services, to provide "recommendations for how to better include minority communities in health care research" (Harris & Fletcher, 1997).

The children related to those involved in this event, it is easy to imagine, would carry the widest spectrum of feelings about the role of science in history, everything from pride, to guilt, to neutrality, to rage. These children come to class with more than just prior knowledge about nature; they come with prior knowledge about science itself, and it is "all" these children that I want science educators to imagine teaching.

Social Justice and Equity Within the Current Framework
The *Standards* express a concern for equity most clearly through the slogan of "science for all students" (p.20). The authors signal the importance of this programmatic objective by locating it just after their rationales for science education in general (scientific literacy and economic skill). The emphasis on "science for all" seems to be a response to criticisms that science has been an elite subject or at least that the quality of science education has been stratified so that a small elite get a high–quality education while others get an inadequate preparation in the sciences. In short, the authors wish to socialize the quality of instruction, i.e., not make the quality of science education a form of social capital that divides a set of have and have-nots. Hence, in a chapter on the *Standard's* principles, it is explained that "the *Standards* do not exacerbate the differences in opportunities to learn that currently exist between advantaged and disadvantaged students" (p.20).

Science Education Through Situated Knowledge 131

The have and have-nots are of two types. On the one hand, there are the historically weighty differences of race, class, and gender. On the other the standards are also concerned with the differential education of those who imagine themselves in, or are being slotted for, technoscientific labor markets: scientists, doctors, nurses, technicians, programmers, etc., and the rest of the populace, the technoscientific laity. These two are also to receive the same quality education, where quality is precisely what the rest of the document defines, i.e., what are the qualities of a worthy science education. As the *Standards* reiterate, "science is for all students...regardless of age, sex, cultural or ethnic background, disabilities, aspirations, or interest and motivation in science..." (p.20).

> *They informed us that we could get free medicine. A doctor would come down from the North every three months... Really, I would say we were forced into it. We could not get health care. We were poor. We could not get anybody in the city to help us in the country.*—Herman Shaw, experiment survivor, Baltimore Sun; 8/24/97

The *Standards* approaches these two sets of distinctions through various strategies. First, differences are made to disappear by fiat, i.e., by proclamation. As Alberto Rodriquez (1997) has noted, the *Standards* render invisible the substantial role that gender, race, sexuality, and class have played in the sciences and science education, while decreeing such social differences obsolete. Second, varying interests in the sciences are dealt with by including content components that are traditionally part of science, technology, and society (STS) curricula: history of science, science and society, and philosophy of science, in order to "develop an understanding of the human dimension of science" (p.200).

> *To our African-American citizens, I am sorry that your Federal Government orchestrated a study so clearly racist.*—President Clinton, Washington Post; 5/17/97

In Macon County, Alabama, this human dimension has historically taken a particular racist form. African American men, presumed to be poor and semiliterate, were simply deemed less than human, not worthy of medical care, beyond empathy. As Jones points out, when public criticism began to appear, many medical professionals leapt to the PHS's defense (pp.281–2), leaving it to journalists to expose the inhumanity of the experiment.

The Locus of the Human Dimension

Some space is granted to these discussions in the *Standards*. Specifically there is a content standard called "Science in Personal and Social Perspectives" which outline what the authors see as the ethical standards of science. At the high school level they declare that "peer review, truthful reporting about the methods and outcomes of investigations, and making public the results of work" guide science (pp.200–201), and they assure us that censure follows for those who do not follow these guidelines. At the middle school level the *Standards* speak directly to our case: "Scientists and engineers have ethical codes requiring that human subjects involved with research be fully informed about risks and benefits associated with the research before the individuals choose to participate" (p.169). What narrows the possibility of discussion is, first, that the ethical and practical norms of science are never discussed as contradictory, e.g., the rights of subjects with the need for objectivity. Second, while the *Standards* claim that "the effect of science on society is neither entirely beneficial nor entirely detrimental," no detriments or risks are ever listed, although the same page assures the reader that "science and technology have contributed enormously to economic growth and productivity among societies and groups within societies." Benefits are listed; risks are left to the imagination. The net effect is that ethical conundrums (rather than so-called "codes") are lost in a generally supportive if not adulatory attitude toward technoscience and technoscientists.

> *A Norwegian study of infected sailors, completed about 1910, followed symptoms in whites. The Tuskegee study was meant to build on that work and see if African-Americans were affected differently. It was characterized as a noble experiment. The public health service was proud of it.*—Baltimore Sun; 8/24/97

The *Standards'* worshipful attitude toward scientists is revealed in their first principles, which insist that science education should "reflect the intellectual and cultural traditions that characterize the practice of contemporary science" (p.21). The principle defines this practice/ tradition as "a way of knowing that is characterized by empirical criteria, logical argument, and skeptical review" (p.21, reiterated on p.201). While seemingly commonsensical, this is a mythologizing of the scientific enterprise.

To single out one dimension of this mythology (others will be analyzed in the next section), the image of science as empirical, logical,

Science Education Through Situated Knowledge

and skeptical contradicts twenty years of research on the practice of science itself. Historians, sociologists, and anthropologists have attended quite closely to what scientists do, and they have consistently found that science as it is practiced is characterized by extra-logical rhetoric to convince allies and opponents alike (Latour, 1987), as well as by the use of aesthetics, reputation, and careerism in the selection of theory and the interpretation of data (Kuhn, 1970), the reliance on metaphysical constructs to make sense of empirical data (Hacking, 1983, which is cited in the *Standards*, though it is not at all clear to me how this volume influenced their formation), and passionate attachment to well worn-theory in the face of anomalous data (Kuhn). These scholars, among others, have shown that such characteristics are not incidental but play a critical role in the selection and elaboration of knowledge about "nature" within the sciences. In other words, science, as a practice, rather than as an ideal, is far in excess of the cartoon presented by the *Standards*.

The *Standards* make a nod to the literature I am referencing. Ian Hacking's volume *Representing and Intervening*, for instance, is cited as a resource. Furthermore, the first component of Content Standard G is "Science as a human endeavor." The *Standards* concede in a final point in the "guide" to this standard that scientists "are influenced by societal, cultural, and personal beliefs and ways of viewing the world. Science is not separate from society but rather science is a part of society" (p.201). However, this claim is given no supporting details, no elaboration of the mechanisms that science as practice, theory, and knowledge function as a part of the societal whole. No reference is made of the classic studies of how sexist or popular metaphors have guided scientific theory and practice (Haraway, 1989; Martin, 1991; 1994). Or, to use the words of philosopher Joseph Rouse (1993), no exploration is carried out of science as an "open" and "heterogeneous" system.

Furthermore, this point is lost in a sea of statements idealizing science as empirical, logical, and skeptical (these are elaborated at length on the same page) and driven by an ethics of "peer review, truthful reporting about methods and outcomes of investigations, and making public the results of work" (pp.200–201). (N.b., this "ethic" also fails to bear up under scrutiny. Anthropologists have noted that biology has increasingly been privatized reducing a certain amount of both procedures and knowledge to corporate secrets. Physics has for a much longer time been indebted to military funding, and some portion of physics research has remained hidden behind the shield of military secrecy (on strategies for dealing with the military in physics in Japan and the U.S., see Traweek, 1996a)

> *The audience [at President Clinton's apology] ...included Mary Harper, who as a young student nurse worked on the Tuskegee experiment; she said she knew an injustice was being done but felt powerless to stop it.*—The Washington Post; 5/17/97

My argument is not that science does not involve empiricism, logic, peer review, skepticism, etc., but that these do not form an essential core. The extent of their role and their very meanings depend on the field, the location, and time in history. By metonymically substituting these ideals for the daily practice of science we put it on a pedestal and heroicize scientists. This pedestal is easy to locate historically, as the ideals presented are precisely those held in philosophy and sociology prior to Kuhn's *Structure of Scientific Revolutions*.[1] As Hacking notes, prior to Kuhn's claims, philosophy (and I would add sociology) "had long made a mummy of science" (p.1). (On the sociological side, see Bloor [1976]; on the mummification of science in science education, see Kuhn, chapter XI; Kelly et al. [1993])

> *More than 120 of the 399 men died of the disease or its complications... Dozens of wives were infected. Some gave birth to infected children.*—Reported by Sandy Banisky, Baltimore Sun; 8/24/1997

For the teacher of our test case, the *Standards* offer few resources. While the relatives of the PHS staff might find comfort in the pristine view of science as logical, skeptical, etc., those related to the subjects are asked to empathize, if not honor, people who treated their relatives as animals.[2] This is an invitation to rebellion, passive or active. In other words, if the only position we are allowed to imagine ourselves in, to empathize with, is that of people who have been our oppressors, we then have only the choice of becoming our own oppressors or its opposite: complete rejection. These students' lack of possibilities reflects the lack of possibilities offered by the *Standards* as they "reflect" pre-Kuhnian romances of science in the guise of "the practice of contemporary science" (p.21).

Roads out of Scientific Monotheism

If these students are to be given a science education, then some plurality of relationships with science must be provided. This plurality has multiple dimensions. First there is the plurality of what constitutes scientific practice. As I have already noted, citing Kuhn, Haraway, and Hacking, what counts as method varies widely between speculative fields and applied fields, ethological field studies and laboratory microbiology, genetics and high energy physics, etc. There is no core here, family resemblance maybe, but the fields which study nature and which we call sciences are proliferating, not converging, and as the proliferation proceeds such resemblances will be harder and harder to find (on the genealogy of conceptualizing the sciences as unified, see Galison & Stump, 1996).

A second plurality that emerges in the empirical, discursive, and philosophical studies of the sciences concerns the locus of scientific knowledge itself. Starting with Bruno Latour's (1983; 1987) criticism/development of the early ethnographies of laboratories (e.g., Latour & Woolgar, 1986/1979),[3] the network of science has been shown to be increasingly expanded. Scientists remain dependent on their political and economic sponsors and often on a broad populace who remain connected to science through infrastructure such as the electric power grid, techno-medicine, and diverse consumer products. Latour poignantly asks, "who is really doing science?" (1987, p.157) and implies that it is in essence people in all of these places, for the discovery of a neutron or of a new antibiotic could not proceed at all without the whole of this complicated network. "Those who are really doing science are not all at the bench, on the contrary, there are people at the bench because many more are doing science elsewhere" (p.162). The image of the scientist in the laboratory or the field is replaced with a proliferation of jobs, institutions, and processes, the lone hero with a network. The object of this wildly expanded conception of science for those studying it is not to replace one node of the network with another (laboratories with legislative bodies, for instance) but to track the movement, translation, and enrollment of people in ever–expanding construction of links and nodes.

Yet this network vanishes in the *Standards*. Scientific practice in every photo, and in every example, reduces to a second idealization of science as the world outside the laboratory or field station is rendered invisible. The necessary janitors, politicians, partners/spouses, secretaries, lobbyists, manufacturers, financiers, and consumers are gone. The laboratory exists in some Platonic world outside the human fray (even though the *Standards* acknowledge this "human influence" in passing).

These nodes, however, are not homogeneous zones, areas marked by consensus or (perhaps) even reified things unto themselves. Houses, laboratories, medical clinics, schools, and field stations are places where differently "historied" people and materials pass through or come to dwell and struggle for different meanings of the sciences and of nature(s) both generally and in their particulars. This constitutes a third multiplicity in science. Anthropologists and others have tracked, for example, in careful ethnographic and discourse studies the conflict over meanings between medical practitioners and patients, around AIDS (Epstein, 1996; Treichler, 1987), cancer (Patterson, 1987; Stacey, 1997), childbirth (Davis-Floyd, 1992), and fetal monitoring (Rapp, 1990; 1993). Furthermore, nationality, gender, and race fragment any simple unity we might find in massifying terms like "scientists," "subjects," or "patients." Traweek (1988) and Haraway (1989) both compare the knowledges, practices, and cultures of American and Japanese scientists. Hess (1994) has tracked complicated transnational countercultures in cancer research. Martin (1990), Rapp (1990), and Davis-Floyd (1992) reveal a wide variety of patient "subcultures" with very different interpretations of their bodies, pregnancy, and technological interventions.

> *All I knew was that they just kept saying I had bad blood, they never mentioned syphilis to me, not even once.—Charles Pollard, experiment survivor,* Bad Blood

As others have argued, science cannot be simply represented as a consensus on knowledge about nature (Apple, 1990). Locations, such as schools or laboratories, might be better thought of as the contexts, or generators of multiple conversations and knowledges generated by differently located people, i.e., people, or even human and nonhuman actors as recent theory would have it, located on various sides of various boundaries. Scientists and patients in this version of science are separated by credentials (perhaps), technology (even as simple as a stethoscope), and roles, but people in all positions are generators of knowledge; the scientists have no monopoly. (Subject/patient knowledge takes one concrete form in the job zine/journal *Guinea Pig Zero*, which serves the human subject community. The zine recounts the history of human subjects and reviews different research establishments for the quality of treatment, benefits, etc.)

> *Middle-school students are generally aware of science-technology-society issues from the media, but their awareness is fraught with misunderstandings. Teachers should begin developing student understanding with concrete personal examples that avoid an exclusive focus on problems.—*The Standards, p.168

Knowledge, therefore, takes on a localized quality. It is tied, first of all, to the different nodes of technoscience, and second, it is bounded by

Science Education Through Situated Knowledge 137

the different zones constructed, often hierarchically, within these nodes (e.g., scientist vs. subject vs. administrative staff vs. clerical staff). These pairs form the contexts for knowledge production—though I must reiterate that such productions cannot be thought of as unities or final claims, only as ongoing conversations or arguments. There is no Archimedian point in the node or in the network that can account or consume all the knowledges produced within. Knowledge in science, in this telling, is by definition partial. To mark its boundedness, Haraway (1991b), following Suchman (1987), describes it as "situated," and contrasts it with the oft-universalizing tone of scientific discourse; she notes that "infinite vision is an illusion, a god-trick" (p.189). Situated knowledge is the inverse and is for Haraway foundational: "The moral is simple: only partial perspective promises objective vision. This is an objective vision that initiates, rather than closes off, the problem of responsibility for the generativity of all visual practices" (p.190). Scientists, as they spin out explanations of nature, are positioned and bounded in multiple ways. First, they are bound by the instruments they use and through which nature is translated into data. Second, they are situated in their disciplines, which inevitably prefer particular tropes (metaphors, models, etc.) over others and which bear particular histories including varying involvements in the dynamics of class, race, gender, sexuality, etc., vis-a-vis other human and nonhuman subjects/objects, including those they study, those that clean up after them, those that pay their way, and those that hope to exploit their findings, etc. Finally, they are positioned relative to other people in other nodes in this web called science.

Some Americans, including children of some of the Tuskegee study participants, believe the experiment was more sinister than the government has acknowledged, that doctors took healthy men and injected them with syphilis bacteria.—Reported by Sandy Banisky, Baltimore Sun; 8 /24/1997

This picture of science as a complex network fraught with fractures, nonresolving multiplicities, and power/hierarchies is a far cry from the NSES image of logical, skeptical, ethical scientists as the end-all and be-all of scientific production. This level of complexity and multiplicity, I argue, provides one route to a science education for all the students in my *Gedenken*[4] class.

Multifocal Science Education

I would argue that by changing the image of science from one of deified scientists to the rhizomatic proliferation of knowledges by multitudes of located people/actors (Martin, 1998), science education itself must change. How do we move from a multifocal view of science to a science education friendly to that image of knowledge production? One simple way that this could happen is by making the multiplicities themselves the subject of science education. Students would not merely study the work, ethics, and knowledge of scientists but also of patients/subjects, service workers (janitors and secretaries), and consumers, as merely a small subset of the possibilities. But each of these locations should also be understood as containing multiple, perhaps even contesting, knowledges. Students could take the opportunity of playing each of these roles, shifting lenses from producer, to servicer, consumer, or subject of science. In addition, they could "play" the science studies scholar, examining in addition to the competing claims each of these locations generate, the way these claims are circulated and given legitimacy.

I am not advocating for substituting lab knowledge with knowledge of some other location. This is not an argument for science teachers using *Guinea Pig Zero* instead of Watson's *Double Helix* (1968) or Sharon Traweek's (1988) accounts of high energy physics rather than Weinberg's *Dreams of a Final Theory* (1992). But it is also not the opposite either, a privileging of the scientists' voices over those of their subjects or others who are participant/observers to lab knowledge production, which at least in the U.S. in some sense or other, is everyone (as I write this a small piece of my skin is sitting in a laboratory being analyzed for cancer. In this sense I am literally already in the lab).

> *1995—In a shocking study, a White House panel asks 1800 randomly selected patients in hospital and clinic waiting rooms around the country whether or not they were participating in medical research. 412 of these patients turned out to be research subjects, including 99 who said they were not and never had been.... Close to 25% of the people surveyed who were participating in medical research did not realize it. According to the study, about 20% of these unknowing participants were involved in experiments that posed more than a "minimal" health risk.—Joseph Gabriel,* Guinea Pig Zero, *#5*

How would such a modified curriculum address the concerns of how to teach our *Gedenken* classroom? First, in contesting the pure logical, skeptical, ethical, and empirical image of the sciences, the heartless attitudes of the Tuskegee scientists themselves could become a legitimate topic of the classroom discussion (given the narrow definition of scientific ethics in NSES, Tuskegee could only be seen as bad science, not science doing bad). This discussion could explore the links between objectivity and objectification (Keller, 1985). In exploring the limits of objectification, the class might begin to investigate not only the development of scientific understandings of diseases like syphilis, but might contrast this with the history of human subjects used to produce this knowledge (as explored, for instance, in the pages of the "zine" *Guinea Pig Zero*).

Perhaps more importantly, legitimacy can be given to the multiple accounts, interests, and desires of the subjects of the Tuskegee experiment themselves, *in addition* to those of the doctors, scientists, and their assistants, since the voices of scientists are positioned as only one set among many others in the multifocal perspective. Including multiple voices opens the possibility of multiple roles being played, multiple identifications, and finally, multiple relationships with the institutions of science production and dissemination. Students are not asked to identify with the PHS scientists *over* those of their subject relatives or the other way around. Instead, students with sympathies in every direction are asked to understand the conditions that left these men vulnerable to exploitation, on the one hand, and permitted the doctors to view these men as less than human on the other. Furthermore, multiple knowledges are open for discussion here: medical accounts of syphilis and minority accounts of medicine. Neither is invalid here; both have qualities we might characterize as reliable.

In this "hypothetical" classroom, at least, a multifocal approach to science education truly does provide a *science education for all*. It does so by insisting that students come to comprehend not science as a monolith, either in its claims, practices, or ethics but as a web or rhizome of contesting and dialoging claims and practices. Rather than reinforcing the image of science as a citadel of rationality and logic, this approach to science education insists on nomadicity (Boundas, 1993) or world traveling (Lugones, 1990) as a means of understanding the multiplicity of rationalities and logics that gather around those practices, products, and knowledges called the sciences.

Conclusion

Science education has long been about cultivating a worshipful attitude about the sciences. In the *Standards*, for instance, it is not enough to understand science, but education must be about "the richness and excitement of knowing about and understanding the natural world" (p.13), where knowing and understanding are to be carried out in the language and through the practices of science as defined by the *Standards*. The goal here is not just to comprehend but to feel excitement about science, to leave us in wonder at its explanatory and material power (for a discussion of the discourse of wonder and science, see Weinstein, 1997). To accomplish this it draws on theological discourses which formerly placed wonder in the realm of the divine (Daston, 1991). Through this authority, the professions of science seek not just to substitute their claims about nature for nature itself but also gather to themselves an ultimate authority about the natural.

The anthropological and cultural studies of science have sought to displace this monopolization. The point here is not, as Haraway has pointed out, apostasy (Haraway, 1991a), the rejection of science, but is instead about localizing and specifying scientific knowledge. In arguing for the partiality of all knowledge claims (even these) she says:

> The kind of partiality I'm talking about is resolutely antitranscendentalist and antimonotheist....Any transcendentalist move is deadly; it produces death through the fear of it. These holistic transcendentalist moves promise a way out of history, a way of participating in the god trick. A way of denying mortality. (cited in Penley & Ross, 1991, p.16)

This religious resonance between scientific universalism and monotheism is also criticized by Traweek, who asks

> Why should there be only one way to think well, only one way to have fun with our minds? Why is mental monogamy required? Are we still fighting about monotheism....Does thinking without singularities mean we cannot think carefully about ourselves, other human beings, and our phenomenal world? Not only are we doing it, we already know how to be playful and graceful as we think, dance, and sing about ourselves, the other humans, and our world. (Traweek, 1996b, p.148)

This multiplicity (as we find it in post-Kuhnian nature narrated by the sciences, in scientific practices and ethics, in the nodes of the sciences, and the zones within those nodes) should guide us in what Paul Hurd (1994) has called postmodern science education. What I have tried to outline here is one way of making science education loyal to this multiplicity.

Finally, it might be asked if this imagining of science education has any relevance outside our hypothetical classroom. I would argue that all classrooms, certainly those I have known at the high school level, have students with complicated relations to those knowledges, practices, and professions identified with the sciences. All students come with prior knowledge and experiences, and therefore attitudes, about the politics, ethics, and cultures of the sciences. In all classrooms students bring diverse narratives of salvation and suffering at the hands of doctors, teachers, and other people who operate under the signifier of science. Given this, the curriculum model sketched above offers enormous possibilities for helping students comprehend and feel powerful vis-a-vis scientific knowledges. This is an education for all students, i.e., for all the roles that we/they come to have in relation to the plurality of knowledge called the sciences. This is a science education that prioritizes the "all" in "science for all students."

> *[Senator Edward Kennedy] said, "You were a guinea pig." I said, "No, Senator, we were all grown men. I think they used us for guinea hogs."— Herman Shaw, experi-ment survivor,* Baltimore Sun; 8/24/97

In conclusion, the partiality of my own account should be acknowledged. This represents only one of many routes toward a more equitable science education. First, the findings of the anthropology and cultural studies of science are loudly contested by portions of the academic and scientific communities (Gross & Levitt, 1994; Ross, 1996). Obviously, if my premises are dismissed, some other route to providing equity in science education will have to be crafted. But even accepting this literature would not determine the curriculum practice I have suggested in this paper. Susskind and Finkel (1995) have developed pedagogies to address exactly the same concerns about equity and exclusion that I have outlined in this paper.[5] Their approach involves combining feminist theory and Freirian pedagogy to teach students to be citizen-scientist-activists rather than exploring the multiplicities that run through the body scientific. The approach I have outlined here, however, adds to the repertoire of equity and socially sensitive science educations by stressing the heterogeneity and openness of the sciences and respecting the variety of attitudes toward the sciences that students bring to the science education classroom.

Endnotes

1. The history of standards are worded vaguely enough to be read as either supporting or contradicting Kuhn's analysis of paradigms in science.
2. While it is acknowledged that the Tuskegee scientists acted unethically, the reasons their actions were sanctioned for forty years by

the medical-scientific community remains mysterious, since officially—i.e., according to the *Standards*—such behavior is against the norms of science. It should be noted that ultimately it was congressional hearings that terminated the experiment, not the censure of the scientific community.

3. The criticism was that laboratory studies "were at risk" of being internalist, since the researchers had often limited themselves to watching scientists only in the laboratory setting.

4. *Gedenken* refers to a thought experiment. These were made popular by Einstein in his reasoning about relativity

5. See also Rodriguez (1997) and Eisenhart, Finkel, and Marion (1996) for further analysis of the *Standards*.

References

Apple, M. W. (1990). *Ideology and curriculum* (2nd ed.). New York: Routledge.

Bloor, D. (1976). *Knowledge and social imagery*. London: Routledge & Kegan Paul.

Boundas, C. V. (Ed.). (1993). *The Deleuze reader*. New York: Columbia University.

Calabrese, B., A., & Osborne, M. (1997, April). *Liberatory science education outside of schools: Playing with identities and borders*. Paper presented at the meeting of the NARST (National Association of Research on Science Teaching), Chicago, IL.

Daston, L. (1991). Marvelous facts and miraculous evidence in early modern Europe. *Critical Inquiry, 18*, 93–124.

Davis-Floyd, R. E. (1992). The technocratic body and the organic body: Cultural models for women's birth choices. In D. J. Hess & L. L. Layne (Eds.), *The anthropology of science and technology*. Greenwich, CT: JAI Press.

Eisenhart, M., Finkel, E., & Marion, S. F. (1996). Creating the conditions for scientific literacy: A re-examination. *American Educational Research Journal, 33*(2), 261–295.

Epstein, S. (1996). *Impure science: AIDS activism, and the politics of knowledge*. Berkeley, CA: University of California.

Galison, P. L., & Stump, D. J. (1996). *The disunity of science: boundaries, contexts, and power*. Palo Alto, CA: Stanford University.

Gross, P. R., & Levitt, N. (1994). *Higher superstition: The academic left and its quarrels with science*. Baltimore: The Johns Hopkins University.

Hacking, I. (1983). *Representing and intervening: Introductory topics in the philosophy of natural science.* Cambridge: Cambridge University.

Haraway, D. J. (1989). *Primate visions: Gender, race and nature in the world of modern science.* New York: Routledge.

Haraway, D. J. (1991a). A cyborg manifesto: Science, technology, and socialist-feminism in the late twentieth century. In D. J. Haraway (Ed.), *Simians, cyborgs, and women: The reinvention of nature* (pp. 149–182). New York: Routledge.

Haraway, D. J. (1991b). Situated knowledges: The science question in feminism and the privilege of partial perspective. In D. J. Haraway (Ed.) *Simians, cyborgs, and women: The reinvention of nature* (pp. 183–201). New York: Routledge.

Harris, J. F., & Fletcher, M. A. (1997, May 17). Six decades later, an apology. *The Washington Post,* p. A01.

Hess, D. (1994). Alternative cancer research and the construction of national, legal, and scientific boundaries. *American Anthropological Association,* Atlanta, GA: AAA.

Hurd, P. D. (1994). New minds for a new age: Prologue to modernizing the science curriculum. *Science Education, 78*(1), 103–116.

Jones, J. (1993). The Tuskegee syphilis experiment: "A moral astigmatism." In S. Harding (Ed.), *The racial economy of science.* Bloomington: Indiana University.

Jones, J. H., & Tuskegee Institute. (1981). *Bad blood: The Tuskegee syphilis experiment.* New York, London: Free Press; Collier Macmillan Publishers.

Keller, E. F. (1985). *Reflections on gender and science.* New Haven: Yale.

Kelly, G. J., Carlsen, W. S., & Cunningham, C. M. (1993). Science education in sociocultural context: Perspectives from the sociology of science. *Science Education, 77*(2), 207-220.

Kuhn, T. S. (1970). *The structure of scientific revolutions* (2nd ed.). Chicago: University of Chicago.

Latour, B. (1983). Give me a laboratory and I will raise the world. In K. Knorr-Cetina & M. Mulkay (Eds.), *Science observed: Perspectives on the social study of science.* Los Angeles: Sage.

Latour, B. (1987). *Science in action.* Cambridge, MA: Harvard University.

Latour, B., & Woolgar, S. (1979/1986). *Laboratory life: The construction of scientific facts.* Princeton: Princeton University.

Lugones, M. (1990). Playfulness, "world"-travelling, and loving perception. In G. Anzaldúa (Ed.), *Making face, making soul, haciendo caras: Creative and critical perspectives by women of color*. San Francisco: Aunt Lute Foundation Books.

Martin, E. (1990). Science and women's bodies: Forms of anthropological knowledge. In M. Jacobus, E. F. Keller, & S. Shuttleworth (Eds.), *Body/politics: Women and the discourses of science*. New York: Routledge.

Martin, E. (1991). The egg and the sperm: How science has constructed a romance based on stereotypical male-female roles. *Signs, 16*(3), 485.

Martin, E. (1994). *Flexible bodies: Tracking immunity in American culture—From the days of polio to the age of AIDS*. Boston: Beacon Press.

Martin, E. (1998). Anthropology and the cultural study of science. *Science, Technology, and Human Values, 23*(1), 24–44.

National Research Council. (1996). *National science education standards*. Washington, DC: National Academy Press.

Patterson, J. T. (1987). *The dread disease: Cancer and modern American culture*. Cambridge, MA: Harvard University.

Penley, C., & Ross, A. (1991). Cyborgs at large: Interview with Donna Haraway. In C. Penley & A. Ross (Eds.), *Technoculture* (pp. 1–20). Minneapolis: University of Minnesota.

Rapp, R. (1990). Constructing amniocentesis: Maternal and medical discourses. In F. Ginsburg & A. L. Tsing (Eds.), *Uncertain terms: Negotiating gender in American culture* (pp. 28–42). Boston: Beacon.

Rapp, R. (1993, November). *Real time is prime time: Ultrasound fetal images, medical maternal discourse, and popular culture*. Paper presented at the meeting of the American Anthropological Association, Washington, DC.

Rodriguez, A. J. (1997). The dangerous discourse of invisibility: A critique of the National Resource Council's National Science Education Standards. *Journal of Research in Science Teaching, 34*(1), 19–37.

Ross, A. (Ed.). (1996). *Science wars*. Durham, NC: Duke University.

Rouse, J. (1993). What are cultural studies of scientific knowledge? *Configurations, 1*(1), 1-22.

Stacey, J. (1997). *Teratologies: A cultural study of cancer*. London; New York: Routledge.

Suchman, L. A. (1987). *Plans and situated actions: The problem of human-machine communication.* Cambridge: Cambridge University.

Susskind, Y., & Finkel, E. A. (1995, April). *Encouraging the socially responsible use of science: Examples of curricular reform.* Paper presented at the meeting of the American Educational Research Association, San Francisco: CA.

Traweek, S. (1988). *Beamtimes and lifetimes: The world of high energy physicists.* Cambridge, MA: Harvard University.

Traweek, S. (1996a). Kokusaika, Gaiatsu, and Bachigai. In L. Nader (Ed.) *Naked science: Anthropological inquiry into boundaries, power, and knowledge.* New York: Routledge.

Traweek, S. (1996b). Unity, dyads, triads, quads, and complexity: Cultural choreographies of science. In A. Ross (Ed.), *Science wars.* Durham, NC: Duke University.

Treichler, P. A. (1987). AIDS, homophobia, and biomedical discourse: An epidemic of signification. *Cultural Studies, 1*(3), 263–305.

Watson, J. D. (1968). *The double helix: A personal account of the discovery of the structure of DNA* (1st ed.). New York: Atheneum.

Weinberg, S. (1992). *Dreams of a final theory.* New York: Pantheon Books.

Weinstein, M. (1997, April). *Dellsifying science.* Paper presented at the meeting of the Central States Anthropological Association, Madison: WI.

Chapter Six
Genre Analysis as a Way of Understanding Pedagogy in Mathematics Education
Susan Gerofsky

Contemporary ideas about genre can be traced to the work of literary analyst and linguist Mikhail Bakhtin. Bakhtin's notion of "speech genres" (1986) includes a very wide range of spoken and written forms, from brief verbal utterances to letters, plays, and novels. Bakhtin's concept of genre has been adopted by a number of disciplines, including film studies, folklore studies, and linguistics, as a useful "fuzzy set" category which can be characterized, but not defined, by a constellation of co-occurring features.

In this paper, it is argued that genre analysis is a form of discourse analysis that can provide useful and sometimes surprising new perspectives in understanding teaching and learning. An analysis of the written and spoken "texts" of schooling, drawing on linguistic and literary analytic methods, can highlight relationships between educational genres and other culturally recognizable forms. The study of such related genres may allow researchers to uncover hidden cultural meanings, assumptions, and intentions inherent in the generic forms of schooling.

Two examples from my research will be used to illustrate the application of genre analysis in mathematics education. The first is a study of mathematical word problems which analyzes the genre features of word problems and suggests a relationship between word problems, parables, and riddles. The second study analyzes the language of initial calculus lectures at a university and finds generic similarities with the language of the conjurer, the salesperson, or nurse. Both studies question the messages, intentional and unintentional, carried by generic forms in mathematics education.

What Is Meant by Genre ?

Many writers (for example, Buscombe, 1970; Sobchack, 1986; Palmer, 1991) credit Aristotle with outlining the first notions of genre in western culture. Concepts of genre in the second half of this century, however, have developed in relation to Bakhtin's notion of "speech genres." The work of Bakhtin, the Russian literary critic and philosopher of language, has been tremendously influential in contemporary Anglo–American literary and linguistic studies, even though much of his work has only been available in English since the mid-1970s.

Bakhtin's concept of genre takes in a very wide range of language phenomena, oral and written, from both "high" and "low" culture. Bakhtin stresses "the extreme heterogeneity of speech genres, oral and written" and cites as examples of speech genres everything from "short rejoinders of everyday dialogue" and "everyday narration" to "business documents...the diverse forms of scientific statements and all literary genres (from the proverb to the multivolume novel)" (pp.60–61).

Bakhtin's basic unit of analysis is the utterance, a unit of language in use and in context. Bakhtin stresses the dialogic nature of all oral and written language—even monologue is framed as a form of dialogue—and sees the quality of addressivity, of addressing a known or imagined other, as key to the understanding of utterances. Bakhtin sees the listener or audience not as a passive partner in dialogue but as a force constantly shaping the utterance of the speaking or writing subject through the listener's/ reader's forthcoming or anticipated response. In Bakhtin's theory, speech and writing can never be removed from a context of addressivity, of dialogue. There is always a ground for the linguistic figure, although the same figure can be reframed in terms of different grounds.

For Bakhtin, there is no possibility of utterances that exist outside of genre. Bakhtin writes, "all our utterances have definite and relatively stable typical forms of construction of the whole" (p.78). Like Molière's Monsieur Jourdain, who had no idea he was speaking in prose, Bakhtin writes, we "speak in diverse genres without suspecting that they exist" (p.78).

What Does Genre Study Offer to Mathematics Educators?: Ideas from Genre Studies in Other Fields

Bakhtin's concept of genre has been adopted and developed in literary, linguistic, film, and folklore studies. Theorists in other cultural fields have raised issues which are also useful for a discussion of educational genres.

The film theorist Andrew Tudor, in a foundational essay on film genre (Tudor, 1973), puts forth the problem of the "empiricist dilemma" in identifying genres—in this case, the "western" genre in popular films:

Genre Analysis

> To take a genre such as a western, analyze it, and list its principal characteristics is to beg the question that we must first isolate the body of films that are westerns. But such films can only be isolated according to their principal characteristics. We are caught in a circle that first requires that the films be isolated, for which purposes a criterion is necessary, but the criterion is, in turn, meant to emerge from the empirically established common characteristics of the films. (p.5)

Tudor's suggestion for a way out of this dilemma is to avoid establishing a priori criteria and instead to "lean on a common cultural consensus as to what constitutes a western and then go on to analyze it in detail" (p.5). This solution presupposes a high degree of shared culture and the likelihood of shared patterns of recognition and meanings that are part of that culture. Tudor suggests that "from a very early age most of us have built up a picture of a western. We feel that we know a western when we see one...." Similarly, mathematical word problems are familiar enough in our culture to be immediately recognizable in popular culture contexts as varied as a TV episode of "The Simpsons," a newspaper cartoon, an advertisement, or a joke. To call a film a western, or a math question a word problem, is more than to define it as sharing certain principal characteristics with other westerns or word problems; it also suggests that the item would be more or less universally recognized in our culture as belonging to that particular genre category. Tudor writes, "Genre notions...are not critics' classifications made for special purposes; they are sets of cultural conventions. Genre is what we collectively believe it to be" (p.7).

The form and addressivity of school mathematics word problems are recognizable nearly universally among most people who have attended school mathematics classes. The extraordinarily long history and widespread educational use of word problems (they have been used continuously around the world for over 4,000 years starting in ancient Babylonian and Egyptian cultures) seem to suggest that they have existed as a cultural form since the development of the earliest writing systems.

A Question of Intentions: Conscious Intentions vs. the Rhetorical Constraints of Genre

Another film theorist, Thomas Sobchack (1975), emphasizes genre films' conservatism in matters of form. He writes that

> There is always a definite sense of beginning, middle, and end, of closure, and of a frame. The film begins with 'Once upon a time...' and ends only after all

> the strings have been neatly tied, all major conflicts resolved. It is a closed world. There is little room in the genre film for ambiguity anywhere. (pp.105–106)

A similar conservatism of form, and sense of closure, is evident both in mathematical word problems and in initial calculus lectures. (A more detailed analysis of these forms follows in later sections of this chapter.)

Sobchack, in reference to their quality of mimesis, writes that genre films

> ...are made in imitation not of life but of other films. True, there must be the first instance in a series or cycle, yet most cases of the first examples of various film genres can be traced to literary sources, primarily pulp literature...And once the initial film is made, it has entered the pool of common knowledge known by filmmaker and film audience alike. Imitations and descendants—the long line of 'sons of,' 'brides of,' and 'the return of'—begin. (p.104)

There is an interesting parallel here with the question of word problem as genre. Word problems in mathematics education are written "in imitation, not of life, but of other word problems." I think this is a necessary realization for those mathematics educators who sincerely try to write word problems that more closely imitate life, or relate to the lived experiences of their students. I contend that the genre form itself speaks directly to the students in terms of what is given them and expected of them; readers dive into the world created by the genre as into a familiar (warm or ice-cold) bath. Word problems imitate and recall other word problems, not our lived lives.

Genres recall and are related to other genres. The term "intertextuality" from reader response theory and hermeneutics describes the ways a reader's awareness of other texts comes to bear on the reader's interpretation of a text at hand, which in turn affects the reader's awareness of text in general. Similarly, inter-genre awareness comes to bear on our interpretation of genres as a whole and on particular generic texts, which exist in a cultural universe mediated by other established cultural types and forms.

Rhetorician Karen Jamieson (1975), in a paper on rhetorical genre, shows that generic forms carry with them a history of culturally and linguistically coded intentions. The use of a generic form may bring with it intentions that are not exactly the same as those of the current writer or speaker. Jamieson gives the example of a papal encyclical which borrows from the rhetorical forms of a Roman imperial edict. She argues that contemporary readers interpret a papal encyclical and its Latin prose style as pompous, turgid, and overbearing in part because the encyclical is written like a message from the Roman emperor.

Word problems, as a very old generic form, also carry with them intentions which act as a rhetorical constraint to the intentions of contemporary mathematics educators. Rifts exist between the conscious, stated intentions of educators and the force embedded in the word problem genre itself. Contemporary writers and teachers may consciously intend to refer to their students' lived lives, to "real-life" situations when they offer word problems to their students. The genre itself may constrain or subvert these conscious intentions through intentions carried by the generic form—the intention not to refer in a straightforward, literal way to the "things" talked about in the word problem, but in a coded and ambiguous way to both those real-life things and, more strongly, to the "things" or objects of a world of mathematical concepts different from the world of everyday experience.

For this reason, educators' stated intentions with regard to the use of word problems may be read as a justification or alibi, in terms acceptable within our culture, for a form which doesn't necessarily fit that culture. Mathematical word problems, in a form very similar to those written for contemporary math textbooks, have a continuous written history that goes back more than four millennia, to the ancient Babylonians, Egyptians, and Greeks, as well as to medieval China, India, Europe, and the Middle East. The generic form carries with it intentions that are throwbacks to cultural norms of mathematical education from earlier times and other places. Word problems are interesting as artifacts in the archaeology of mathematics education, sedimented as they are with meanings from other worlds. They are much more than simply straightforward applications of mathematics to real life.

The rhetorical scholar Carolyn Miller (Miller, 1984, quoted in Swales, 1990) writes,

> To consider as potential genres such homely discourse as the letter of recommendation, the user manual, the progress report...[and I would add here "the mathematical word problem" and other educational genres] is not to trivialize the study of genres; it is to take seriously the rhetoric in which we are immersed and the situations in which we find ourselves. (p.155)

Miller argues that

> what we learn when we learn a genre is not just a pattern of forms or even a method of achieving our own ends. We learn, more importantly, what ends we may have. (p.165)

This speaks most cogently to the acculturating power of forms within mathematics education and education generally—that generic forms, which are familiar and recognizable but often existing below the threshold of our conscious recognition, define us in our relationship to our worlds. Through genre forms, we learn what may be asked and what is beyond question in our culture, what we may aspire to and what is outlandish or forbidden. Questioning and exerting pressure on genre is a way of questioning particular unspoken boundaries of culture, and this is especially important when working in a culture like that of school mathematics, with a strong tradition of conservatism and exclusion.

Case Study: Mathematical Word Problems as Genre

St. Peter talking to a man at the gates of heaven: "OK, now listen up. Nobody gets in here without answering the following question: A train leaves Philadelphia at 1:00 p.m. It's traveling at 65 miles per hour. Another train leaves Denver at 4:00....Say, you need some paper?"
—Gary Larson, "Far Side" cartoon

If a joke leaves New York at 11:30, what time does it reach Vancouver?
"—Ad in the Vancouver Skytrain for local television rebroadcast of the David Letterman Show."

A man leaves Albuquerque heading east at 60 miles an hour. At the same time, another man leaves Nashville heading west at 65 miles an hour. Which man is closer to Nashville when they meet?
—Joke told by an eight-year-old to her mother and father, overheard in a restaurant in Thunder Bay, Ontario, Canada.

Mathematical word problems are instantly recognizable to most people in our culture who have been to school, so much so that they can be used without comment or explanation in jokes, cartoons, and advertising. There are many aspects of their form which are familiar, including sentence structure, the sequence of story elements, data, questions, and the kind of stories and imagery typically used. Aspects of the uses of word problems in mathematics classes are also familiar to most people. These problems are to be translated into mathematics (arithmetic, algebra, geometry) and the mathematical question is to be solved using taught methods to find the right answer. Enough data are given in the statement of the problem itself, and the problem solver may not demand more data, although often "red herrings" may be thrown in to the problem to trick inexpert solvers. Adults' emotional reactions on being presented with a word problem are usually intense: intensely negative (in most cases) or intensely positive. Gary Larson plays on our shared recognition of word problem phobia in the cartoon text quoted above.

In my study of word problems in mathematics education as a genre, I use analytic theory from linguistics, literary criticism, and mathematics education. I consider the question, "what *are* word problems?," and try to answer it by "taking a walk" around the word problem genre as an object to see it from many points of view, including the linguistic, the historical, and the pedagogical. A consideration of genre leads to questions of addressivity and intention. In this case, intention includes educators' conscious, stated intentions when writing and assigning word problems but also includes intentions carried by the word problem genre form itself and students' uptake of their teachers' intentions, of which educators may not be aware. I suggest that an exploration of genre in mathematics teaching and learning can be a source for innovation and renewal in mathematics education practices.

Characteristic Features of the Word Problem Genre
In a previous paper (Gerofsky, 1996), I looked closely at the constellation of features that characterize word problems as a class or genre. Moving among points of view grounded in linguistics, literary criticism, and mathematics education, I found that the following features typify mathematical word problems:

1. A three-component, sequenced rhetorical structure (a "disposable" story element, followed by data and then a question).
2. "Indeterminate locutionary force"—that is, the referents for the nouns used in word problems are ambiguous. "Deixis" (pointing with words) in word problems is problematic because the nouns used do not refer in any but the most tangential way to their usual "real-world" referents. Words in mathematical word problems point to some other world than that of our conscious, lived, real-world experience.
3. Strongly imperative "illocutionary force" (i.e., coded intention of writer): "Solve this problem to get the right answer, using only the information given and the mathematical methods I have just taught you!"
4. "No truth value." As with fiction, it is impossible and unproductive to assign the statements in word problems "true/ false" status (Frege's notion of truth value,

adopted into linguistics) because their existential presuppositions are false. However, word problems are very poor quality fiction at best and are deliberately constructed so that their stories are considered interchangeable with a great many mathematically equivalent stories.

5. An anomalous use of verb tense. Word problems combine verb tenses in ways that would be jarring in ordinary speech. This use of verb tense is shown to be "tenseless and non-deictic"—in other words, the strange mixing of verb tenses is another indication that the words in mathematical word problems point to some other world than that of "real life," so that we are not bothered by their lack of consistency when referring to time.

Related Genres

Analysis of the linguistic and rhetorical features of a particular genre may be suggestive of other genres within the culture. This consideration of "intertextuality" of a genre which, in its form and addressivity, recalls other familiar genres, may bring to consciousness the hidden ground and intentions imbedded within the genre. If one genre is recognizably like another, then echoes and resonances among them can inform a reassessment or refiguring of genre-related cultural practices. Rather than being forced into a binary choice of either approving or disapproving of a particular genre (as with contemporary calls either to do away with or reemphasize the use of word problems in mathematics classes), it becomes possible to look at the genre with a sidelong gaze, and consider questions like, "What if we treated word problems as we do parables? How would we use them for teaching then?" In this next section, I will take a sidelong, metaphorical view of word problems along with resonant, related cultural genres, with the implied question, "What could this mean for the ways we use this genre in education?"

Mathematical Word Problems and Parables

"Teaching stories" are familiar in Western religious cultures, in such forms as Biblical parables and Talmudic stories, and in Eastern religious cultures in forms like Sufi teaching tales and parables from Buddhist sutras. In this century, parables have been adapted to nonreligious contexts; for example, writers like Franz Kafka, Søren Kierkegaard, and Flannery O'Connor have taken up the parable for secular philosophical purposes.

Parabolic teaching stories share with word problems a particular, and peculiar, nonreferential use of language. Or perhaps the term "nonreferential" is too absolute. Rather, their language refers only in the most tangential way to real-life referents but primarily points to some "other world"—in the case of parables and other teaching stories, to the

Genre Analysis 155

world of religious, spiritual, or philosophical entities, and in the case of word problems, to the world of mathematical entities. In both cases, seemingly referential language is used to express otherwise inexpressible feelings and ideas through concrete images, which could just as well be substituted by innumerable alternate concrete images, so long as they served to point to the same abstract entities in the primary domain of reference. Parables share with word problems an indeterminate use of verb tense and interchangeability of characters, situations, and scenes for structurally equivalent referents within their proper domain of reference. As with word problems, the assignment of truth or falsity to teaching tales is irrelevant, since the tales are not referring to an everyday world where truth values can be established.

Kafka, in his book *Parables and Paradoxes*, talks about the nondeictic nature of parable in the form of a parable:

> Many complain that the words of the wise are always mere parables and of no use in daily life, which is the only life we have. When the sage says: "Go over," he does not mean that we should cross to some actual place, which we could do anyhow if the labour were worth it; he means some fabulous yonder, something unknown to us, something too that he cannot designate more precisely, and therefore cannot help us here in the very least. All these parables really set out to say merely that the incomprehensible is incomprehensible, and we know that already. But the cares we have to struggle with every day: that is a different matter. (Kafka, 1961, p.11)

Clearly, the parables do not refer in a nonambiguous, straightforward way to the usual referents from our lived experiences. When a parable refers to fishermen, the intention is not to talk about the particularities of fishing. In fact, too great an emphasis on the details of real-life fishermen's work would be a distraction from the point of the parable, which may be to talk about lost souls, spiritual guidance, and so on.

In a similar way, mathematical word problems may purport to talk about real-life situations, but in fact too great an insistence on the contingencies of the experiential situation distracts from the point or intention of the word problem. One such example is cited by Christine Keitel (1989). She recounts observing a lesson in which a teacher wanted to teach ratio and proportion in a practical way. Keitel writes:

> She offered the following question: 'Somebody is going to have his room painted. From the painter's samples he chooses an orange colour which is composed of two tins of red paint and one-and-a half tines of yellow paint per square metre. The walls of his room measure 48 square metres altogether. How many tins of red and yellow are needed to paint the room the same orange as

on the sample?' The problem seemed quire clear and pupils started to calculate using proportional relationships. But there was one boy who said: 'My father is a painter and so I know that, if we just do it by calculating, the colour of the room will not look like the sample. We cannot calculate as we did, it is a wrong method!' In my imagination I foresaw a fascinating discussion starting about the use of simplified mathematical models in social practice and their limited value in more complex problems (here the intensifying effect of the reflection of light), but the teacher answered: 'Sorry, my dear, we are doing ratio and proportion.' (p.7)

When a word problem is offered that appears to deal with painting, for example, it is important not to introduce much real-life contextual knowledge about painting. That is not what the word problem is *about*; it is not what it is "pointing to" with words. The mathematical word problem is *about mathematics* in the same way that the religious parable is about religion, and the same lesson about proportion could be taught through a story problem about mixing cookie dough or cement, about the ratio of girls to boys in a school, or any number of stories that could illustrate proportional thinking.

Word problems are also similar to parables in their nondeictic mixing of verb tenses. Where the use of verb tense in English would be governed by certain norms of grammar and logic in ordinary speech and writing, these norms are suspended for both word problems and parables.

Word problems differ most markedly from teaching tales in their illocutionary force—that is, in the intentions of those writing or offering them to learners. Mathematical word problems are typically used in pedagogy to practice established, recently taught solution methods. In this style of pedagogy there is value placed on quick, correct, economical solution of the problematic, without reference to extraneous knowledge of the world outside the stated problem, without further questioning of the contingencies of the story, and without undue puzzlement, irritation, or contemplation. When a word problem has been solved in mathematics class, the problem is usually discarded.

Unlike word problems in mathematics education, parables, koans, and other teaching tales demand no solution and often ask no direct question. They are not meant to be solved and generally present an insoluble, paradoxical dilemma. Teaching with parables and other teaching tales traditionally involves a discussion of contingencies of the story, of the problems of human life that relate to the story, of the sources of its paradox. Teaching tales are certainly not meant to be disposable exercises. They are made to be held onto, to irritate, to resonate. A good parable will inspire contemplation and will be recalled in times of difficulty as a way of trying to make sense of a seeming impasse.

Word Problems, Riddles, and Puzzles
Even when word problem stories appear to refer to aspects of the "real world," their links to the world of lived experience are ambiguous at best. So why are these rather fanciful stories included at all?

The importance of unlikely and attention-catching stories seems clear when mathematical problems are intended for recreation and entertainment rather than for "serious" educational purposes. David Singmaster, who has done extensive research into the history of recreational mathematics, writes:

> Many problems in recreational mathematics are embellished with a story which is often highly improbable, and this is partly what makes the problem memorable and recreational. (Singmaster, 1996)

Similarly, Jens Høyrup, the Danish historian of mathematics, quotes Hermelink (1978, p.44) in describing recreational mathematics as "problems and riddles which use the language of everyday but do not much care for the circumstances of reality." Høyrup writes,

> "Lack of care" is an understatement...A funny, striking, or even absurd deviation from the circumstances of reality is an essential feature of any recreational problem. It is this deviation from the habitual that causes amazement, and which thus imparts upon the problem its recreational value. (Høyrup, 1994, pp.27–29)

Using "the language of everyday" but "not much caring for the circumstances of reality" is also a very apt characterization of the nonreferential nature of word problems, parables, and riddles as genres. A "lack of care for the circumstances of reality" has been a feature of word problems as early as the Old Babylonian period. Høyrup posits a continuum of nonreferential story problems ranging from the most delightful mathematical recreations to the dullest of school exercises:

> One function of recreational mathematics is that of teaching...This end of the spectrum of recreational mathematics passes imperceptibly into general school mathematics, which in the Bronze Age as now would often be unrealistic in the precision and magnitude of numbers without being funny in any way. Whether funny or not, such problems would be determined from the methods to be trained....Over the whole range from school mathematics to mathematical

riddles, the methods or techniques are thus the basic determinants of development, and problems are constructed that permit one to bring the methods at hand into play. (Høyrup, 1994, pp.27-29)

Riddles, folktales, recreational mathematics, and pedagogic mathematical word problems appear to have developed and spread in similar patterns. In many cases, the boundaries between riddles, folktales, and word problems are determined more by the context of their use than by their form or content. For example, the familiar children's riddle in English, "As I was going to St. Ives," has been traced back to a problem in the Rhind papyrus from ancient Egypt and is related to a problem from Sun Tzu in ancient China. The well-known "crossing the river" problem (as well as the "hundred birds" problem and a number of other widespread problem types) has been reported as a source of village riddling contests in the Atlas Mountains of Morocco and in locations in eastern and southern Africa. In an interview I conducted recently, an Iranian-born mathematician told me about being "riddled" as a boy with a story problem about weighing flour, which he later saw as a specific instance of a more general theorem in number theory.

Høyrup relates the very wide distribution and longevity of a number of famous mathematical word problems to an oral tradition of recreational problem riddles transmitted by merchants along the Silk Route:

Like other riddles, recreational mathematics belongs to the domain of oral literature. Recreational problems can thus be compared to folktales. The distribution of the "Silk Route group" of problems is also fairly similar to the distribution of the "Eurasian folktale," which extends "from Ireland to India."...Recreational problems belong to a specific subculture—the subculture of those people who are able to grasp them. The most mobile members of this group were, of course, the merchants, who moved relatively freely or had contacts even where communication was otherwise scarce (mathematical problems appear to have diffused into China well before Buddhism). (Høyrup, 1994, pp.34-35)

Is there a distinction between riddles, recreational puzzles, and school word problems? Since the same problems can be found contextualized in all three settings, the difference seems to lie mainly in the intentions surrounding the problem. Riddles are contextualized in a setting of pleasurable social interaction. They can be part of a process of building social solidarity and, simultaneously, a source of competition, as in the village riddling contests. Although riddles have been collected in written form, their primary use is in oral culture, and good riddlers can draw from a large memorized repertoire upon which a certain degree of improvisation is possible.

In contrast, word problems in school mathematics are traditionally assigned as a sort of bitter medicine that will make you better. In North American mathematics textbooks, they usually come at the end of a series of "easier" numerically or algebraically stated problems related to a mathematical concept introduced in the preceding chapter. The word problems represent a final test of students' competence in recognizing problem types related to that chapter and translating those problems into tractable diagrams and equations which can be solved using taught algorithmic methods. School word problems are not social events or part of an oral culture. They are ideally meant to be solved silently, individually, using pencil and paper. Students are certainly not encouraged to memorize a repertoire of word problems for later enjoyment. On the contrary, once solved, they are generally discarded by teacher and student.

Riddles often retain strong links to folktales and parables and other teaching tales in their invocation of paradox and ambiguity, through their use of puns, hyperbole, nonsense, etc. Like parables and word problems, they point to two worlds at once—the world of their literal referents and another world invoked by word play or unexpected associations and structures. In riddles and parables, this ambiguity is embraced as essential to the enjoyment and philosophical import of the genre.

Contemporary writers of mathematical word problems, on the other hand, work hard to make their problems realistic, relevant, and unambiguous. In this pursuit of singleness of meaning and relevancy, they are stymied by the genre's history and form, which carry with them the intention to create paradox and at best a shifting relationship to everyday reality.

Implications for Teaching

What if we treated word problems in mathematics classes as if they were parables, or riddles? Would this alter our intentions as mathematics educators or our students' perceptions of those intentions? For example, if word problems were not considered disposable exercises (as they often are now) but as parables worthy of longer and deeper contemplation, we might spend a week considering a single word problem in all its considerations and implications. Students might be asked to comb older textbook or even historical sources for word problems pointing to the same mathematical structures as the "parabolic" story under consideration. They might consider changing certain features of the word problem while holding others unchanged, and seeing whether this altered the mathematical relationships pointed to in the problem. They might try to project real-life situations in which recalling the word problem "parable" might be instructive, or helpful, or comforting.

And what if word problems were considered as riddles? First of all, a playful and perhaps competitive spirit would be invoked. Word play and double meanings would be welcomed rather than banned. A pleasurable, oral culture approach to a recreational use of word problems would take the place of our present, very serious approach to evaluation of student written knowledge.

Perhaps the simple suggestion that mathematical word problems be considered *as parables* or *as riddles*—the shift to the "as if" point of view that characterizes play and drama—may begin to engender a shift in thinking and in educational practice.

Case Study: Initial Calculus Lectures as Genre

In another recent study of genre in mathematics education, I looked at first-year university calculus lectures as a genre. I compared a number of taped lectures by four mathematics professors at Simon Fraser University in Burnaby, British Columbia, Canada (pseudonyms Brown, White, Green, and Black). Using clusters of linguistic features, I found parallels between the "initial calculus lecture" genre and several other culturally significant speech genres.

Some prominent features of the "initial calculus lecture" genre included unusual uses of the first person plural pronoun (we, us, our), extensive use of rhetorical questions and tag questions, the attribution of questions or opinions to the audience and lecturers "answering" these unasked questions or objections, and the structuring of the lecture as a chain of inexorable logic that could lead to no conclusions but the ones given. I was struck by the similarity of some of these features to the language of persuasion, particularly a "hard sell" sales pitch, and of other features to infant-directed language or "baby talk."

In a "hard sell," the salesperson's job is to forge an inexorable chain of logic that leads to one conclusion: that the prospect would be crazy not to buy what the seller has to offer. Typical tools used in the hard sell include demonstrations (think of the vacuum cleaner salesman who pours dirt on his prospect's best carpet), making claims for the marvelous qualities of the product for sale, voicing possible objections or questions the prospect might have and then answering them with prepared responses, identifying oneself with the prospect's disbelief ("I didn't believe it either at first !"), and above all, creating such an unstoppable stream of talk that the prospect has no chance to consider any objections or questions other than those already suggested by the salesperson before the seller has moved in to "close the sale." I suggest that many of these features are typical of the math lecture genre, particularly in calculus lectures addressing students in their first year of postsecondary education.

These features are also typical of other genres that involve "selling" or convincing—religious proselytizing, political speechmaking, "hard sell" advertising in all media. What is being "sold" is not always a tangible commodity; it may be a belief, an opinion, an idea. I think it is in this light that I see the similarity between a math lecture and these other forms of persuasion.

Like other "persuasion artists," the math lecturer is often making a "cold call"—trying to convince an often unwilling and unprepared audience of the truth and efficacy of certain mathematical beliefs. The lecturer is the representative within the lecture hall of the whole field of mathematics—"mathematics personified." The lecturer's problem, as seen through this lens, is to use the lecture hour in the most efficacious way to engage students' interest in the wares that are "for sale" (in this case, mathematical ideas), to convince them of the truth of the arguments the lecturer is to present, and to persuade students to accept those arguments so as to be able to function as if they were self-evident (and use the mathematics involved in problem-solving, exercises, and exams). If generic parallels may be used as evidence, then this has been seen as a "hard sell."

> Goffman (1981) defines lectures in general as follows: 'A lecture is an institutionalized extended holding of the floor....Constituent statements presumably take their warrant from their role in attesting to the truth, truth appearing as something to be cultivated and developed from a distance, coolly, as an end in itself' (p. 165).

The lecturer's monopoly on the right to speak (or to decide who will speak), the presumption that truth exists and that the lecturer can represent and deliver it to an audience are features of all lectures; it is my proposition that they correspond in some ways to the salesperson's extended holding of the floor during the "pitch." The lecture genre, whatever its subject content, is already a mode of persuasive talk that tries to "sell" its audience on both the truth of the ideas presented, and, implicitly, the authority and status of both the lecturer and the sponsoring institution as purveyors of truth and knowledge.

Within the "persuasive" framework of the lecture, math lecturers use other discourse techniques that further the impression of lecturer as pitchman. All the lecturers in my study used the rhetorical question to some extent, and most used the old sales trick of raising "fake" questions and objections on behalf of the listener and then answering them with prepared responses as if the audience's real questions had then been addressed. Here are a few examples:

> You might think it's just square root of one plus one over x plus one over x squared, but it isn't, it's the negative of that. Let's see why it's the negative of that rather than just the square root. (White)

> Somebody's going to say, "Why not have infinity on *both* the input *and* the output sides?" Well, why not? (White)

> So you've got a big numerator over a big denominator and at first glance you might say, "I haven't the faintest idea what's happening". But you should because this x fourth is more ambitious than the x cubed. I think if you think about it for a moment you can sense that the x fourth is sooner or later going to completely outstrip the x cubed and it's going to be more or less like one over x. (White)

> And now you've got two things and they're getting large and they're working at cross purposes and now it's not clear what's going to happen and maybe somebody's going to say, "It's getting large positive, this is getting large negative, you're going to get zero." No, you're not. To see what you get this time, again we pull our famous stunt of dividing top and bottom by the same thing, so here goes. (White)

> Now you want to know,"Where is this function maximum?" And the first thing you might think to do, well, I know that the turning points or the points when that slope is zero is when something happens on the function. OK? (Green)

> And then you might want to say, "OK, then I'll take the second derivative and use that to tell whether or not I have a maximum or a minimum." Now that works fine a lot of the time. But a lot of the time it doesn't help you. So we're not going to do the second derivative. We're just going to do the first derivative, then we're going to think about what's happening to this function. (Green)

Schmidt and Kess (1986), in their study of linguistic persuasion techniques in television advertising and televangelism, cite experimental studies on the effect of rhetorical questions on persuasion. It seem that the effect varies with the listener's degree of involvement:

> In the case of rhetorical questions...high-involvement subjects actually found the use of this type of question distracting, making them less sensitive to the quality of argumentation than they were when the same arguments were presented as assertions. Low-involvement subjects, however, showed greater sensitivity to the quality of argumentation when the message contained rhetorical questions than when it did not. This effect was explained by the fact that rhetorical questions essentially ask the hearer to think about the topic, thereby increasing the level of message-based thought for low-involvement subjects, a processing strategy which they would not normally have engaged in to the same degree. (Schmidt & Kess, 1986, p. 31)

These findings fit well with the argument that math lecturers use "hard-sell" persuasion techniques; the whole idea of a "hard sell" is that the prospect is hard to sell to, an unwilling listener, a "low-involvement subject." Rhetorical questions are effective in persuading such a listener although counter-effective in persuading a listener who has greater interest and commitment to the topic.

The particular style of rhetorical questioning noted in the examples above, which we could call "fake dialogue," has been noted in other types of lectures and other discourse styles as well. Leith and Myerson (1989), discussing a university English lecturer, note that "in effect he holds a dialogue with imagined voices articulating the opposed position, and he assumes that some of his students at least some of the time may identify with those voices" (Leith & Myerson, 1989, p. 16). In a later discussion of a political speech, they find a similar example of "fake dialogue": "The speaker holds a dialogue not only with the audience, not only with opposite but absent voices, but also with the previous speaker at the conference" (Leith & Myerson 1989, p. 23).

Goffman (1981), in his discussion of radio talk, finds radio announcers engaging in "fake dialogue":

> So announcers must not only watch the birdie; they must talk to it. Under these circumstances, it is understandable that they will often slip into a simulation of talking with it. Thus, after a suitable pause, an announcer can verbally respond to what he can assume is the response his prior statement evoked, his prior statement itself having been selected as one to which a particular response was only to be expected. Or, by switching voices, he himself can reply to his own statement and then respond to the reply, thereby shifting from monologue to the enactment of dialogue. In both cases the timing characteristics of dialogue are simulated.(Goffman, 1981, p.241)

A further example of "fake dialogue" has been noted by Ervin-Tripp & Strage (1985) in interactions between parents and preverbal children in which parents "interpret burps, or single- or two-word constructions as if children have elaborate intentions, confirming, expanding and elaborating them" (Ervin-Tripp & Strage, 1985, p. 73). All these examples of "fake dialogue" have in common the fact that they are addressed by a speaker to an audience that cannot respond, either because they are not present (in the case of the radio announcer or the political speaker), because they are incapable of responding (in the case of infant-directed talk), or because they are socially constrained from responding (as are the audiences for the university lecturer or the speaker at the political conference). The pedagogical question then becomes, should students learning mathematics be treated as if they cannot or must not respond ?

The lecturers made extensive use of tag questions ("Right?", "OK?", etc.) to elicit audience consent. These were used to elicit both agreement with a statement made by the lecturer ("It's positive...OK?") and permission to move on to the next section of the lecture. ("Very stringent requirement. Alright? So now I'd like to do this case.") Here are some examples:

> Now another thing you want to make sure to understand is that, hey look, two ordered pairs are the same if and only if their first co-ordinates have to be the same and their second co-ordinates have to be the same. Right? Very stringent requirement. Alright? So now I'd like to do this case. (Brown)

> When theta is equal to zero I get this point. When theta is equal to pi over two you get this point. The in-between points, just take my word, OK ? (Brown)

> The farmer has to, with whatever fencing he has to, has—perhaps we should say she—wants to build a rectangular pen. OK ? (Green)

> When x is negative the square root of x is in fact minus x because it's the positive square root, OK ? And x is negative. Think about it. (White)

This use of tag questions is typical of the persuasive talk of both the salesperson and the conjurer (of which more later). Tag questions have several effects in terms of persuasion. When an audience is addressed with such tag questions, and if there is no loud and vocal audience protest, the assumption is that there is tacit agreement with the point the speaker is making or that permission has been granted for the speaker to move on to another topic or example. Presumably the speaker is "reading" the faces of the crowd and seeing approval there, in the subtle

Genre Analysis 165

form of slight nods of the head, hints of smiles, approving eye contact. In fact the members of the audience are seated so that they can't see one another's faces and are taking it on trust that the speaker has received approval from the majority of the audience. Short of a minor rebellion in the ranks, the audience is carried along on the lecturer's unstoppable stream of talk. The implication is that any rational person, having signalled acceptance of one link in the lecturer's chain of logic, will be led on inexorably to the next link, and the next, and finally reach the one inevitable conclusion.

There are other phrases that seem to be taken straight out of a sales pitch that appear in the examples above and elsewhere in the tapes. Brown's "just take my word" is nearly the same as the salesman's "trust me"; White's "think about it" (with no pause allowed for people to really think about it) has much the same ring.

It may be appropriate to make a further comparison here with the ironically persuasive discourse of the conjurer, ironic because we are persuaded at the same time as we know that we are being tricked. (In some ways this is not very different from being persuaded by a very skillful salesperson whose "tricks of the trade" are very evident—we are persuaded in spite of our better judgment.) Brown's and White's lectures bear particular similarities to conjurers' discourse. At several points Brown slowly and laboriously builds up audience approval of very elementary and obvious statements (like the conjurer's "test" of his materials with a volunteer from the audience: "Is this a solid table? Pass your hand underneath it. No holes in this scarf?," etc.), then, like the conjurer, performs a series of rapid moves and, presto, something unexpected has appeared.

> Now another thing you want to make sure to understand is that, hey look, two ordered pairs are the same if and only if their first co-ordinates have to be the same and their second coordinates have to be the same. Right? Very stringent requirement...and a very obvious point—SG] Alright? So now I'd like to do this case—I'd like to represent this parabola very simply like this....[followed by rapid and unfamiliar algebraic formulation] (Brown)

Both White and Brown hint at the mechanics of illusion—disappearances, levitation, things mysteriously expanding or shrinking—which are metaphors common to the magician and the mathematician. Both make marvelous claims or set challenges to the audience in the course of their lectures, then carry out demonstrations to prove their claims are true.

So what I'm saying when I say "the limit of one over x squared is plus infinity" is, really, I'm saying this: You give me a high horizontal line, and I'll show you how to get the graph to stay above that line. (White)

You give me any number; no matter how big a number you give me, I can show you how to make one over x squared even bigger than that and stay bigger than that. (White)

Now in this case, you see, the difficulty we faced here disappears! (Brown)

Now I claim this is a function! You see...[demonstration follows] (Brown)

To see what you get this time, again we pull our famous stunt of dividing top and bottom by the same thing, so here goes. (White)

Who Are "We" Anyway?

David Pimm (1987) has discussed the use of the first person plural pronoun (we, us, our) in the mathematics classroom. I would like to apply some elements of his analysis to the nonstandard uses of *we*, which appear frequently in initial calculus lectures. I also draw some conclusions about the use of *we* in persuasive discourse.

Pimm outlines four familiar contexts in which the noninclusive *we* (a "we" that does not really include both the speaker and the listener) is used. These are baby talk ("We're getting you dressed, aren't we"), hospitals and doctors' surgeries ("Now we are going to take our temperature"), formal discourse ("In our attempt to analyze addition...we are thus led to the idea...") and school use ("Susan, we never bite our friends") (Pimm, 1987, p.67). (Ervin-Tripp & Strage [1985] note that features of the baby talk register "are also found in speech to pets, dolls, hospital patients, lovers, and the elderly" and hypothesize that "they may indicate affection or protectiveness of the weak" [Ervin-Tripp & Strage, 1985, p.72]). Some conclusions Pimm draws are that this atypical use of *we* may indicate a hearer who is unable to respond (a use that is functionally related to the "fake dialogue" discussed earlier) or may signal a relationship of power and dependence and carry an implicit message of condescension, indicate social convention being conveyed, and/or point at the existence of an "in-group" of mathematicians who carry authority within the field of mathematics. He also mentions the use of *we* in textbooks as attempting "to enrol the (at least tacit) acquiescence of the reader in what is being expressed" (Pimm, 1987, p.68), a use parallel to that of tag questions discussed above.

Genre Analysis 167

Pimm questions the effect of this *we* on the personal or the self within mathematics, and he cites several examples where there is confusion about the referent for *we*. I have found a number of similar examples in the math lectures in this study:

> Now what we want to talk about today is parametric equations. (Brown)
>
> But before we do anything we have to figure out ways to describe the curve in a more satisfactory manner. (Brown)
>
> But we will try to show you people a better way of describing a curve. (Brown)
>
> So now, here is the idea that will save us. (Brown)
>
> And we say this limit is plus infinity. Which is just a way of speaking. All it means is this. You give me any number no matter how big a number you give me, I can show you how to make one over x squared bigger than that... (White)

There is a good deal of confusion in these examples about exactly who is included in *we*. Consider the "we" who want(s) to talk about parametric equations: is it just the lecturer, the lecturer and the students (unlikely!), the lecturer and the rest of the math department, or mathematicians in general? In Brown's next two quotes there is a contradiction in the referent. The first statement ("But before we do anything...") seems to include the listeners, perhaps in a condescending way as a nonresponding audience, in the group of those who "have to figure out ways to describe the curve in a more satisfactory manner"; but in the following quote ("But we will try to show you people a better way of describing a curve"), the listeners have been separated out as "you people" and are excluded from the "we." What, then, is the referent? As in the first example, it seems possible that it could be the lecturer, the lecturer and the math department, or the depersonalized authority of mathematics in general. Similarly, White's "And we say this limit is plus infinity" must exclude the listeners; otherwise there would be no need to explain the statement further. This "we" seems to refer more clearly to mathematicians in general (an in-group which excludes first-year math students). Brown's "So now, here is the idea that will save us" is a bit more difficult to interpret. Does it mean "Here is the idea that will save you, the students, when you try to work on this problem"? Or does that us refer to "me, you and mathematicians in general"? (There is also the problem of interpretation of the word "save" here. In its ordinary meanings, it carries a hint of evangelism (being saved by an idea or a belief—

"eternally saved"?) There is also the connotation from medieval science of "saving the appearances"—finding analytic processes that allow a fit between theory and observed phenomena. (Or it may just mean that a conventional practice may save students from having to work out a more complicated solution to a class of problems.)

Many of the examples of nonstandard use of "we" fall into two patterns: *let's*, and *we* + *do*. The use of *let's* has an air of invitation about it, as well as the implication of condescending baby talk register mentioned earlier. The use of *we* + *do* seems to fall within Pimm's category of "a common fear in mathematics of involving, and hence exposing, the self" (Pimm, 1987, p.70). In these examples, *we* + *do* usually seems to mean *I* + *do*, and the confusion this creates is shown by White's midstream changes of pronoun ("we still need to do something....And what I decided to do..."). Green's "It's easy to see..." is worth remarking. Elsewhere in her lecture she uses other encouraging phrases ("It's easy, right?" "Now we're almost done," "You don't even have to think any more—well, a little bit"), which again recall the somewhat condescending, power-dependence relationship typical of baby talk, doctor-patient talk, and so on. Tag questions, rhetorical questions, nonstandard use of "we," and "making encouraging noises" co-occur in many of these examples, as they do in the baby talk register:

> Let's just take a look. When t is equal to zero, what is my x? X is equal to minus one, y equal to minus one, that's this point here. So let's say t equal to one, let's see what it looks like. (Brown)
>
> Let's draw a picture over here. (Green)
>
> In order to see this — initially we have to do it by brutal force. (Brown)
>
> Now what happens on the other side of 75, when x is between 75 and 150? It's easy to see that the slope is now negative. So if we look at our picture, our tangents go in the other direction. (Green)
> OK, now we're going to get into lots of trouble I think if we start doing this because all these things—we're not going to be able to control the variables very well. (Green)
>
> Now let's see if we can see what's going on and if we can't, we're going to have to do more to it. (White)
>
> And we still need to do something with it. And what I decided to do with it was to divide numerator and denominator by x, but carefully. (White)

Genre Analysis 169

There is a link between the nonstandard use of "we" and the language of persuasion in the mathematics lecture. Charles Ferguson, the linguist who first defined the parameters of baby talk, writes, "The BT [baby talk] register often serves the purpose of coaxing or persuading.... This 'wheedle' use, as Read calls it, can even be extended to objects or to unknown addresses. Read offers, among other examples, Alexander Woolcott's amusing habit of using BT in addressing his dice at backgammon.... Another example of extension is its use in certain forms of advertising" (Ferguson, 1977, p.231). Again, an analysis of discourse features of the mathematics lecture genre leads to an interpretation of the genre as persuasive talk.

One final point I would like to consider in this discussion is the role of metaphor and neologisms (new coinages) in the genre of the mathematics lecture. Pimm has noted that metaphor "is central to the development of the mathematics register, and an understanding of the processes involved is essential to anyone attempting to make sense of mathematical speech or writing" (Pimm, 1987, p.109). The lecturers in this study all used metaphor extensively, along with metaphorically based neologisms. The metaphors I found fell into two categories: personification and metaphors of physical movement over time (usually to describe a static graphic image or a numerical sequence):

> As soon as x gets a decent ways away from zero it's safe to divide by it. And if it's approaching minus infinity it won't be long before it leaves zero way behind. (White)

> The limit as x approaches zero from zero's left hand side....(White)

> Now a vertical asymptote is simply a vertical line like in this case the y-axis that the graph snuggles up to because it's having some kind of an infinite limit. And it's alright if it just snuggles up to it on one side, it's still an asymptote. And it doesn't have to snuggle up to both the positive and negative extensions of the line, just the one end of it. (White)

> The feeling about that, now you've got a war between a term that's blowing up and a term that's going very, very negative, and our feeling is I think that this one wins. And that feeling is right because you could write it this way.... (White)

> Inside here, the one over x and the one over x squared also die out but the one survives so you get minus one.... (White)

> So the thing is not blowing up, it's blowing down! (White) [a neologism based on the metaphor of "blowing up" meaning approaching infinity—"blowing down" means approaching negative infinity—SG]

A couple of squares went flying!...By that time everything was OK. (Black)

Metaphor is such an accepted feature of mathematical talk and math teaching that it would be hard to imagine a metaphor-less mathematics. I think it is clear that metaphor and neologisms play a role in describing, by association, new relationships and phenomena (and indeed it has been suggested that all natural languages are made up mostly of extended metaphors). Bergmann describes some metaphors as "fecund" or having "organizing power"(Bergmann, 1982, p.491), and this power can be very useful in mathematics (or sometimes counterproductive when the metaphor is stretched beyond the limits of its organizing power). Having taken all this into consideration, I would like to suggest one other use of metaphor in the context of the mathematics lecture—its use as a tool of persuasion.

Sandell, summarizing research into the persuasive effects of linguistic style, writes,

> The use of metaphors and similar figures has been subjected to some research. Bowers and Osborn (1966) varied the final parts of two different speeches in two versions: one literal and one metaphorically intense.... On both speeches, the effective difference between the versions was significant, the metaphorical ending increasing attitude change in the advocated direction. Perception of competence, trustworthiness, and ingenuity was affected in a very complex way by interactions between alleged speaker, topic, and type of metaphor.... In a recent study, Reinsch (1971) investigated the effects of the metaphor and still another related figure, the simile, in persuasive discourse. Supporting Bowers and Osborn, he found metaphors aiding persuasion to a significant degree, compared to a literal version. As he predicted, the effects of similes was (sic) between those of the metaphor and the literal versions.... (Sandell, 1977, p.77)

And in their study of television advertising and televangelism as persuasive language, Schmidt and Kess identify neologisms as a key feature of persuasive discourse. Lakoff (1982), in a discussion of persuasive discourse, noted that an essential identifying feature of persuasive communication is its quest for novelty of expression. As she puts it, "persuasive discourse wears out; ordinary conversation does not" (Lakoff, 1982, p. 31). Evidence for the role of novelty in persuasive communication was taken from examples of television advertising which were found to exhibit the following types:

> 1) lexical novelty or neologism (e.g., devilicious) 2) morphological or syntactic novelty (e.g., the soup that eats like a meal)...Lakoff accounts for this extensive use of linguistic novelty as follows. First, anything neologistic,

because it violates the Maxim of Manner [one of H.P. Grice's Principles of Conversation–SG], draws attention to itself, and by capturing the hearer's attention increases the impact of the message. Second, through this violation of the Co-operative Principle, neologism forces the hearer to interpret, and therefore to participate in the discourse. According to Lakoff, this active role played by the hearer, in turn, enhances learning and retention, and consequently also persuasion. (Schmidt & Kess, 1986, pp.30–31)

Again there is a strong connection possible between the language of the mathematics lecture and the language of persuasion.

Implications for Teaching

Is all education in fact a process of persuasion? And what is the role of mathematical "proofs" in the persuasive process? (What is being proved by whom, to whom, how, and why?)

Again, the analysis of the initial calculus lecture as genre can serve to draw attention to a hidden ground (the function of persuasion, or the "hard sell," of addressing an audience that cannot respond), to the familiar figure of the university lecture. The resonances among genres here are meant to be suggestive and to offer a new, "sidelong" view of the calculus lecture as genre. Rather than placing practitioners in the binary bind of either praising or condemning initial calculus lectures as they are practiced, I want to suggest possible questions and directions for a recontextualization of the genre. If the calculus lecture is meant to be a sales pitch, how could we, as educators, emphasize or draw attention to this purpose? Sales pitches take a variety of forms in different media—for example, TV and magazine ads, the individual face-to-face approach, telemarketing, and so on. Could any of these be usefully adapted to the teaching of initial calculus courses?

The similarity of the lecture to a conjuring trick has been adapted to the teaching of science to elementary school students. In my home city of Vancouver, the local science museum and a group called Science Alive use a "magic trick" approach to generating student interest in physics and chemistry. (This "science as fun" approach has been thoroughly critiqued in Appelbaum & Clark, in press.) Is this a workable approach to the teaching of calculus? Could an initial interest in calculus in younger students be stimulated by a "math-magician" approach (as is being attempted with science subjects), with a more rigorous explanatory follow-up at a later stage?

On the other hand, what if we didn't want to take a hard-sell approach to the teaching of calculus? If this were the case, initial calculus classes would take a different form. What could this look and sound like? Perhaps different generic resonances could provide mathematics educators with other metaphors for teaching and learning.

How Has genre Analysis Affected My Own Teaching Practices?
Beyond specific classroom activities in the high school where I teach, there is a change in the tone of teaching that has come as part of my awareness of pedagogical genre. For example, I do not see myself as a salesperson for mathematics, nor do I think of the act of teaching as a form of hard-sell persuasion. In my classes, I consciously work against the hard-sell lecture genre (although I cannot say whether some elements of this genre may still unconsciously be present!). I expect that my students will often conceptualize differently from me, and I am interested in learning about their concepts. I expect to be surprised and to continue to broaden my learning about both mathematics and education by listening to students' ideas and concerns. I expect that I will make mistakes, not know things, and behave foolishly from time to time. Working with these expectations, I try to muster as much humility and collegiality as possible in my teaching. I try to treat students with the respect due to fellow learners in a challenging field. In the role of teacher, I see myself as inviting students to enter further into the culture of mathematics, and I try to share my enthusiasm and perplexities with that culture. It is difficult to characterize my own metaphors for classroom time and space, but they *are* different from those of the hard sell.

Until now, I have only used genre analysis as a topic of teacher and student *talk*. Perhaps this could be greatly expanded by engaging in activity and research around genre analysis—for example, when students research a mathematical topic, I could ask them to present their findings using several different genres; or students could work on genre analysis of mathematics textbooks, television shows, or lectures. The focus would be on the different intended and perceived purposes embedded in different genres and on a mastery and enjoyment of genre switching.

In my research on initial calculus lectures, I contrasted the discourse of first-year calculus classes with that of a senior undergraduate mathematics class. There was a marked difference in discourse style, and the tone of the senior class was in many ways parallel to the tone I try to establish in my high school classes.

Professor Black's (a pseudonym) fourth-year undergraduate math classes did not operate in the "math-as-hard-sell" genre. For example, his lectures contained many fewer examples of the noninclusive use of *we* than did the first year course lectures. Black used the personal *I* extensively: "What I'd like to do today...." "My solution...." "I've got compatible states...." "I've got closure." Black based his discourse on a master-apprentice rather than salesperson-prospect imagery. The more intimate tone was that of a seasoned mathematician doing math with a group of junior colleagues, and the "master" was willing to admit he sometimes made mistakes.

Black frequently stepped out of the frame of the lecture to make parenthetical asides to his listeners. Most of these asides were ironic or even self-denigrating comments on his own teaching:

> The example I came up with was wrong. I don't know why I gave it. It was garbage. (Black)
>
> I haven't done this one before. I'm looking forward to checking my solution against yours. (Black)
>
> This is being done just as a teaching technique. (Black)
>
> I think that that point is worth the extra minutes it's taken me. (Black)

Black's tone conveyed an impression of modesty, a more egalitarian footing with the listeners, and an air of consultation with the audience, even within the undergraduate math lecture genre. It stood in marked contrast to the persuasion techniques used in the first-year classes, where students were treated as "low-interest sales prospects" rather than as junior colleagues.

Playing with Genre

The material conditions of postmodernism allow us to be physically in several places at once (on the phone or on the Internet, for example), and so why should we not stand in more than one place intellectually and see things from more than one point of view? Although I may not morally approve of education-as-sales-pitch, why should I not try it on with a sense of irony, a sense of cultural self-awareness, as a costume, as gesture, as art? Playing with genre and pushing it beyond its limits may lead to unexpected insights, may flip a genre into the reverse of its original characteristics (as is suggested by McLuhan and McLuhan, 1988, p.99). Perhaps adopting an extreme generic form in math-as-drama (for example, teaching a lesson in Euclidean geometry in the form of a televangelist's sermon) might bring to consciousness some of the hidden ground of mathematics as a discipline.

References

Appelbaum, P., & Clark, S. (in press). Science! fun?: A critical analysis of design/ content/ evaluation. *Journal of Curriculum Studies*.

Bakhtin, M. M. (1986). *Speech genres and other late essays* (Vern W. McGee, Trans.). Austin: University of Texas.

Bergmann, M. (1982). Metaphorical assertions. In Davis, S. (Ed.). *Pragmatics: A reader*. New York: Oxford University.

Bowers, J. W., & Osborn, M. M. (1966). Attitudinal effects of selected types of concluding metaphors in persuasive speech. *Speech Monographs* 33, 147–155.

Buscombe, E. (1970). The idea of genre in the American cinema. In Grant, B. K. (Ed.). (1986). *Film genre reader*. Austin: University of Texas, 11–25.

Ervin-Tripp, S. & Strage, A. (1985). Parent-child discourse. In Van Dijk, T. A. (Ed.), *Handbook of discourse analysis Vol. 3: Discourse and dialogue*. London: Academic.

Ferguson, C. A. (1977). Baby talk as a simplified register. In Snow, C. E. & Ferguson, C. (Eds.). *Talking to children: Language input and acquisition*. Cambridge: Cambridge University.

Gerofsky, S. (1996). A linguistic and narrative view of word problems in mathematics education. *For the Learning of Mathematics*, 16 (2), 36–45.

Goffman, E. (1981). *Forms of talk*. Philadelphia: University of Pennsylvania.

Grant, B. K. (1986). *Film genre reader*. Austin: University of Texas.

Hermelink, H. (1978). Arabic recreational mathematics as a mirror of age-old cultural relations between Eastern and Western civilizations. In Hassan, A. Y. , Karmi, G. & Namnum, N. (Eds.), *Proceedings of the first international symposium for the history of Arabic science, April 5–12, 1976*, Vol. 2, Papers in European languages. Aleppo, Syria: Institute for the History of Arabic Science, Aleppo University.

Høyrup, J. (1994). *In measure, number, and weight: Studies in mathematics and culture*. Albany: State University of New York.

Jamieson, K. M. (1975). Antecedent genre as historical constraint. *Quarterly Journal of Speech*, 61/4, 406–415.

Kafka, F. (1961). *Parables and paradoxes*. New York: Schocken Books.

Keitel, C. (1989). Mathematical education and technology. *For the Learning of Mathematics*, 9/1, 7–13.

Lakoff, R. T. (1982). Persuasive discourse and ordinary conversation, with examples from advertising. In Tannen, D. (Ed.), *Analyzing*

Discourse: Text and Talk. Washington, D.C.: Georgetown University.

Leith, D., & Myerson, G. (1989). *The power of address: Explorations in rhetoric.* London: Routledge.

McLuhan, M., & McLuhan, E. (1988). *Laws of media.* Toronto: University of Toronto.

Miller, C. R. (1984). Genre as social action. *Quarterly Journal of Speech*, 70: 151–67.

Palmer, J. (1991). *Potboilers: Methods, concepts and case studies in popular fiction.* London: Routledge.

Pimm, D. (1987). *Speaking mathematically: Communication in mathematics classrooms.* London: Routledge.

Reinsch, N. L., Jr. (1971). An investigation of the effects of the metaphor and simile in persuasive discourse. *Speech Monographs*, 38, 142–145.

Sandell, R. (1977). *Linguistic Style and Persuasion.* London: Academic Press.

Schmidt, R., & Kess, J. F. (1986). *Television advertising and televangelism: Discourse analysis of persuasive language.* Amsterdam: John Benjamins B. V.

Singmaster, D. (1996). Personal communication.

Sobchack, T. (1986). Genre films: A classical experience. In Grant, B. K. (Ed.). *Film genre reader* Austin: University of Texas, 102–113.

Swales, J. M. (1990). *Genre analysis.* Cambridge: Cambridge University.

Tudor, A. (1973). Genre. In Grant, B. K. (Ed.). (1986). *Film genre reader.* Austin: University of Texas, 3–10.

Chapter Seven
A Feminist Revisioning of Infinity: Small Speculations on a Large Subject
Elaine V. Howes and Bill Rosenthal

> Since the word "infinite" can stand for a plurality of different things, tell me, which of its various significations do you take in this debate?
> Tullia d'Aragona (1547/1997, p.81)

In(tro)duction

Infinity is an endlessly fascinating topic, particularly to those of us lucky enough to be children. Individually and collectively, our minds teem with conceptions of infinity; you'll discover deliciously disparate ones in such diverse discursive spaces as the cultural lore of both colonial regimes and indigenous cultures, the seminar room of Harvard's mathematics department, and, most delightful to us, the one-up-person's-ship (literally!) of children on the playground ("When I grow up, I'm gonna make infinity dollars!" "Oh yeah? I'm gonna make infinity plus one!!") or their deepest feelings toward those closest to them ("I love you infinity, mommy and daddy"). The assertion that infinity is of universal human interest may be a giant generalization but at most a tiny exaggeration, if one at all.

Infinity is also barely detectable in United States school curricula. We invite you to wonder why this is so—to ponder the paradox that an issue so stirring to human hearts and minds and so central to the intellectual history of our species is for almost all practical and theoretical purposes *topic non grata* in our schools. We have our own ideas, of course, some of which are a touchstone for this chapter. The fact—and we mean, fact—that infinity is all but absent from school is especially notable *vis-a-vis* mathematics, a discipline that has been described as "the science of the infinite" by none other than David Hilbert, a finalist on anyone's list for the preeminent Mathematician of the Twentieth Century. (Since Mathematics is thought by many to be "the servant of the sciences," the exclusion of mathematics from school curricula can be taken to create a wholesale misrepresentation of scientific knowledge and activity.) One explanation, simultaneously resonant and far too facile for infinity's being made unwelcome, is that the very nature of the infinite is so intensely paradoxical and contradictory that it is just too much to handle for even progressive teachers and curricularists. Our opinion is quite the opposite. We believe that infinity's innate, inescapable contradictions make it not merely the ultimate postmodern subject-object and a fitting subject for a volume on postmodern curriculum but a superb article of study for kids from 1 to 92.

We will leave it to others to develop further the too-contradictory theory for infinity's absence from school curricula. Instead—and it is here that we particularize to mathematics and, to a lesser degree, science—we understand (most unscientifically!) this yawning curricular hole as an epiphenomenon of something much larger: A two (plus)-millennia attempt to remove the infinite from Mathematics itself. Those who are interested in details of the conventional tellings of this tale can go to a myriad of hard-core scholarly works in mathematical philosophy, history, and historiography, as well as the many recent popularizations of infinity.[1]

We are choosing to bark up a shakier branch of the tree. To wit: The brunt of what follows is a *feminist interpretation* of the super-strenuous efforts to remove infinity from mathematics (and consequently from school math and science). Be forewarned, please, that our analysis is not conventionally "scholarly." Being neither professional historians, philosophers, sociologists, nor anthropologists, we make no pretense at a taut argument based on unimpeachable historical scholarship, philosophical reasoning, sociological study, or anthropological evidence. Our analytical method is to listen with a feminist receiver to infinity's development—meaning, in essence, that we scan the airwaves for instances of when conceptualizations of the infinite have been vividly related to beliefs and practices about sex and gender, even based upon these. This is not to say that feminism (or women's studies) is less "rigorous" *per se* than other social sciences, only that we are not professionally conversant in any of the other social-scientific disciplines upon which we draw. Thus we offer our findings as speculative and correlational, not deductive and causal. Nevertheless, we find these speculations and correlations to have both emotional resonance for our lives and explanatory power when it comes to the otherwise inexplicable chasm of labeled "infinity" in school mathematics and science.

Since studies of the cultures of the infinite abound in writings accessible to interested laypersons as well as in the intersection of social science with Mathematics and the "hard" sciences, it is sensible to ask, "Why another treatise on infinity?" Good question. To our knowledge, ours is the first attempt to apply feminist theory, deliberately and willfully, to the infinity booth of the Mathematical bazaar. The role of gender in the life history of infinity has been elided. Even if we have missed other feminist analyses of the mathematical infinite, the fact that our search engine has come up empty makes us 99.999...% confident that this contribution will be worth the paper on which it's printed

A Few More Words About the Authors

Now that we have bared our souls on what we are *not*—professionals in philosophy, history, or anthropology, among a horde of other disciplines—here's a little on who we are. For the past few years (5.666...years, to be exact), we have been engaged in a study of conceptions, constructions, and constrictions of mathematical infinity. Beginning with our own fascinations with and memories of the infinite, we set out to develop from a feminist poststructuralist perspective a speculative analysis of the abstract, disembodied Mathematical construction of infinity, utilizing the experiences, perceptions, and critical faculties of one who has been successful at scaling the slippery slopes of academic Mathematics (Bill) and one who has chosen to avoid the climb (Elaine). Let there be no doubt that we both consider ourselves to be *educators* first, foremost, and forever. Bill is secondarily a male lapsed Ph.D. Mathematician; Elaine is a female biologist who has always been enchanted by non-Mathematical conceptions of the infinite yet only marginally comfortable with Mathematics itself.

Before proceeding, some explanatory notes on our typography and jargon. Our use of the uppercase in "Mathematics" and "Mathematical" is anything but incidental or pretentious. Here we follow Alan J. Bishop (1988) in denoting *canonical* Western mathematics as Big-M Mathematics and using the lower case to refer to the plurality of *all* mathematics—a *plural* noun. Jargon-wise (more likely, -unwise), as far as our "feminist poststructuralist perspective," we have already described the "feminist" component as an insistence on seeking out the appearance of sociocultural constructions of sex roles and gendered characteristics in the situation under study. In addition, a feminist perspective connotes an analytical attitude predicated on the premise that knowledge is not independent of the power-laden differentiation of its creators into such mutually exclusive categories as male/female and masculine/feminine. We are "poststructuralist" in our polite but firm refusal of the temptation to house our observations, inferences, and interpretations in any conceptual-theoretical structure that poses as immovable and whose chambers consist of such disjoint and disjointed dualisms as male/female, masculine/feminine, abstract/concrete, art/science, school/ real-life, form/content, empirical/analytic As poststructuralist researchers, we display our lack of conviction that positivist ways of knowing are privileged in providing foundations for knowledge.

A Feminist History of the Infinite

The paths cleared by our explorations have led us back to the storied forefathers (there are no recorded foremothers) of Western thought and their representations of infinity. Contrary to popular historical consensus, mathematical infinity did not begin with Zeno and his paradoxes. Although Mathematicians date the Big Bang of infinity back to Zeno, varied explanations and discussions of its properties and powers were considered by serious philosophers long before he promoted the match race between Achilles and that plodding tortoise.

Unlike contemporary formalized conceptions of Mathematical infinity, these prephilosophical and early philosophical deliberations were explicitly and closely tied to the physical and mystical worlds. Our examination of the gestation and birth of mathematical infinity has led us to hypothesize that Mathematics' separation from the physical and mystical was part and parcel of the separation of matter from mind (the minds of humans, the mind of God) in Western religion, science, and Mathematics. There are clear-to-us connections between the establishment of masculine supremacy and the development of the un-Earthly abstractions of Western Mathematics. Feminist interpretations of the coestablishment of patriarchal political, social, religious, and academic dominance (Dinnerstein, 1991/1976; Eisler, 1987; Griffin, 1978; Lerner, 1986, 1993; Noble, 1992; Stone, 1976) have informed and inspired our examination of gender in mathematics as illuminated in portrayals of the infinite.

The Pre-Pythagoreans

There exists a plethoric panoply of conceptualizations of infinity—Mathematical, mathematical, poetic, and personal—neglected by the lion's share of philosophical and historical scholarship, as well as the popular accounts that are dependent on them. Here we concentrate on pre-Pythagorean ideations.[2] Our spotlight on antiquity before Pythagoras and his crew has revealed strikingly powerful conceptions of the infinite as immanent in the physical world as well as representative of the spiritual.

Historians have knighted the Pythagoreans as the first legitimate

Mathematicians, which is perhaps why authors affiliated with Mathematics have ignored the pre-Pythagorean infinite. However, once upon a pre-Mathematical time, early philosophers conceived (of) the infinite, and they did so in cerebrated and celebrated *physicality*. Foremost among these pre-Pythagoreans was Anaximander (Johnson & Rosenthal, 1996), who philosophized infinity as the "originative substance" of the world (Sweeney, 1972), as "a great unlimited mass enveloping the world [and] that which out of things are formed" (Lear, 1980), and as "a huge, inexhaustible mass...surrounding the world" (Sweeney, 1972).

Things—and we mean the word literally—just don't get any more physical than *that*. In contradistinction to the modern Western insistence on separating the physical and spiritual (and favoring the latter over the former), Anaximander's infinity (*apeiron*) isn't only physical—it's simply *divine*.

> [Anaximander] assumed [*apeiron*] to be of unlimited extent and duration—properties which, when expressed, would be expressed in terms of all-inclusiveness and divine immortality. (Kirk & Raven, 1957; quoted in Sweeney, 1972)

This physical-divine infinite is neither male nor female: In fact, the unbounded, boundaryless substance-spirit whence all is generated is *both* female and male. Not only does *apeiron* give birth to all things, but it "'secretes' the seed out of which emerge the opposing principles whose interaction constitutes the world" (Sweeney, 1972). The pre-Pythagorean infinite was the union, not the disjunction, of woman and man.

The Pythagoreans

Before Pythagorean dualisms, it was all okay. The pre-Pythagoreans were not frightened by infinity, not even wary of it. All indications are that, in their holism, they were at perfect peace with the unlimited, the boundaryless, the inexhaustible, the immortal. It took the Pythagoreans to rend asunder both the cosmos and human existence into *gendered* dualities, in all of which the inequality Male > Female is axiomatic, and elements homologous with the female are to be shunned or feared. The elucidation and solidification of disjoint dualisms appears to be one of the Pythagoreans' prized projects; in their ordering of the world and the cosmos, everything was either/or. As represented by Merrill Ring (1987), the Pythagoreans plopped the infinite into the derogated column that also contains the feminine. Their immensely successful program to subsume *everything* under a system of dualisms resulted in the following scheme, a passing glance at which sheds light on the first Mathematicians' conceptions of number, physicality, spirituality, and gender.

limit	unlimited
odd	even
one	plurality
right	left
male	female
at rest	moving
straight	crooked
light	darkness
good	bad
square	oblong

> Notice which items are lumped with good and which with bad—an important social comment is embodied in the placement of male and female in this schema of conflict. (Ring, 1987, p.56)

We note Ring's recognition of the *gendered* character of the Pythagorean opposites. Annemarie Schimmel (1993) makes the same observation while placing a tad more emphasis on the fact that both the infinite and the female ended up on the wrong side of the tracks:

> The Pythagoreans...went so far as to divide everything in the universe into two categories: the odd numbers belong to the right side, which is associated with the limited, the masculine, the resting, the straight, with lightness and goodness, and, in terms of geometry, with the square, while the even numbers belong to the sphere of the infinite, the unlimited (as they are infinitely divisible), the manifold, the left side, the female, the moving, the crooked, darkness, evil, and, in geometric terms, the rectangle....For Plato, all even numbers were of ill omen, and Hopper states correctly: "As if the feminine numbers were not already sufficiently in disfavor, the stigma of infinity is attached to them, apparently by analogy to the line." (pp.13–14)[3]

Some cultural and disciplinary historians of the infinite have noticed its denigration by the Pythagoreans and their successors. In one of the most widely read accounts of infinity, Rudy Rucker (1982) writes,

> The Greek word for infinity was *[to] apeiron*, which literally means unbounded, but can also mean infinite, indefinite, or undefined. *Apeiron* was a negative, even pejorative, word. The original chaos out of which the world was formed was *apeiron*. An arbitrary crooked line was *apeiron*. A dirty crumpled

handkerchief was *apeiron*. In Aristotle's words, "...being infinite is a privation, not a perfection but the absence of a limit."

There was no place for the *apeiron* in the universe of Pythagoras and Plato (pp.2–3).

Certainly, there wasn't—now that *apeiron* had been reclassified from divine to dark, bad, and female.

Aristotle

Which brings us to Aristotle. Yes, the same Aristotle who thought of women as "mutilated males" (Lerner, 1986, p.207). One day when he decided to take a break from contributing to the oppression of women for at least the next two-and-a-quarter millennia, Aristotle defined two kinds of infinity, the "potential" and the "actual." His description of the potential infinite focused on "showing it to be a potency, something which is always becoming without ever reaching a final form" (Maziarz & Greenwood, 1968, p.174). Thus the potential infinite can and must exist, as it makes change over time possible; in fact, it even makes *time* possible—without the additive quality of the potential infinite, we could not move from one moment into the next. On the other hand, Aristotle argues against the "actual infinity of the universe" (Maziarz & Greenwood, 1968, p.175). The material world is closed in by the heavens, and we cannot imagine anything beyond; thus, we cannot conceive of it as a Mathematical entity, or even describe it utilizing Mathematical thinking.

Thus Aristotle banished the actual infinite from Mathematics; concomitantly, he separated Mathematical calculation from the physical world. Aristotle's depiction of two types of infinity was, at least in part, a response to Zeno's paradoxes. By separating Mathematics from the material, "actual" world (the *real* world?)—completing the job started by the Pythagoreans—Aristotle was able to sidestep Zeno's claim that the world could not be infinitely divisible. Thus the whole concept of an actual infinity—equated with woman, identified with a "dirty crumpled handkerchief"—had to go out the window.

In Aristotle's Wake

We are maintaining that the roots of the Mathematical infinite grew in the same soils from which sprung the dichotomies that became the canonical basis for Western thought, particularly and especially the subordinations of the body to the mind and the feminine to the masculine. These binary oppositions live on to this day, conditioning and structuring every thought we take, every move we make.

Others have interpreted Mathematical constructions of the infinite on the basis of the splitting of the body from the mind. In this vein, Brian Rotman's 1993 *tour de force* of a book is worth lingering on for a moment. Rotman embarks on the tough task of "taking God out of mathematics and putting the body back in." He critiques the ontological basis of Mathematical infinity on the basis of its physical un-realizeability. SomeBODY, a physical being, has to do the counting, and this bodily-ness places physical constraints on how the counting is done. We can't just keep "adding one" *in the exact same way*. If we think of mathematics as something done by a physical subject, physics simply doesn't permit Mathematics to create the infinite the way it has purported to.

Rotman's account of "putting the body back into mathematics" never quite accounts for just *whose* body has earned the privilege. Certainly it is not Anaximander's, nor any of those of the billions of women throughout history who have been excluded from the priesthood that determined what was to be "rigorous" and "legitimate" Mathematics and science (Noble, 1992; Schiebinger, 1989; Wertheim, 1995). Any analysis ignoring the historical fact that, as social groups, men and women have led different lives due to the cultural laws imposed on their bodies, cannot repair the philosophical violence that has riven mind from "the" body. Therefore, Rotman's brave new infinity, brought into being in accordance with sound Mathematical principles, cannot be an infinity with which we are satisfied.

Our research has taught us that the Pythagorean-Aristotle axis transmogrified infinity from ungendered, divine, complete, active, matter-and-form, and physical-and-divine, to feminine, non-divine, incomplete, passive, matter only, and formless physicality. There is a theological tie-in here: The infinite was pre-Pythagoreanally both Goddess and God as well as both Girl and Guy. The Pythagoreans and Aristotle mutated *apeiron* into someTHING nondivine and feminine, while christening the spiritual as masculine. Sometime later—our knowledge here is fragmentary—femininity and infinitude were decoupled, and God Himself, the monotheistic male deity, became the Infinite.

> Philosophers had been wary of the paradoxical nature of the infinite since the Pre-Socratics first began to explore its many contradictory forms. Aristotle's solution was generally followed, which simply rejected the use of completed infinities. Christian theologians were also opposed to the actual infinite; for the

> most part they regarded the idea as a direct challenge to the unique and absolutely infinite nature of God. (Dauben, 1979, p.120)

Aristotle stripped infinity of its dignity. Apparently, latter-day philosophers and religionists wanted to restore the infinite to its throne. All it took was a sex-change operation. No problem.

On a less controversial note, Mathematicians themselves insist that Mathematical reasoning is separate from the physical world. "Formalization" is a desirable goal (Lakatos, 1976; Thomas, 1996); separation from the world of matter is proof that a mathematical object is authentically Mathematical. Plato's ideal forms live on in Mathematics. It is small wonder that a school-mathematics canon still rooted in Platonic absolutism and a school-science ethos founded upon the quantification of science continues to omit almost any mention of the infinite. To those who would base school subjects solely on a ladder-like hierarchy of disciplinary knowledge, infinity cannot be suitably formalized (i.e., tamed) until one has acquired more "Mathematical maturity" than most students will ever have in their worst nightmares.

Summary and Speculations

Those who have written infinity's pre-History (pre- to Zeno, we mean) have passed over gender issues. In their neglect, every time these scholars celebrate the "genius" of Aristotle's argumentation in the infinite (Lear, 1980), they celebrate the debasement of the feminine. As is illustrated by our humble little history and historiography of the infinite, Mathematics has rejected the feminine and thus rejected women. The description of the pre-Pythagorean conception of infinity seems to fit much better with a world in which female and male are both respectable, respected, and necessary—in contrast to the power- and value-laden binary split between female and male that dates back to Pythagorean days. Historians and philosophers of the infinite have generally short-shrifted the pre-Pythagorean version. Not so coincidentally, this rendering preceded the advent of the patriarchal worldview that denigrated infinity by aligning it with the despised feminine. Patriarchy later valorized the infinite by associating it with the male monotheistic deity who was the last god standing.

It is also worth noting that Georg Cantor, the most celebrated infinitist of all, reversed the Aristotelian decree by returning the actual to infinity. Joseph Dauben has devoted a goodly chunk of his career to chronicling Cantor's life and work; we shall not misrepresent him by attempting to condense the Cantorian saga into a sound bite. We do, however, wish to make two observations. The first is Bill's recollection of a remark made some years ago by Dauben (1995). This comment disabused Bill of his naïve misconception that Cantor's infinities—his *actual* infinities—were pure products of his pure mind. Instead, Cantor considered his creations to be derived directly from physical reality—the same (?) feminized physical reality that led Aristotle to banish the infinite from Mathematics and Plato to admonish us not to trust our senses—the same senses that *know* the infinite every second of every day.

We are also much taken by the following comment in Dauben's masterly study of Cantor.

> Finally, mathematicians generally followed the philosophers in avoiding any application of the actual infinite, and their reluctance stemmed from the apparent inconsistencies such concepts always seemed to introduce. (1979, p.120)

We hypothesize that "their reluctance" had and has something to do with male fear of the feminine. A thread runs throughout the history of Western Mathematics of fear, or at least irritation, concerning concepts of the infinite.

> The hostility of mathematicians toward infinity began in the fifth century B.C.[E.] when Zeno of Elea, a student of Parmenides, formulated the well-known paradox of Achilles and the tortoise. (Moore, 1995, p.112)

A second thread, first woven by the Pythagoreans and Aristotle, binds the cloth of Western philosophy and psychology (Neumann, 1994). We are speaking of fear and loathing of the feminine—often in the form of *naming* as feminine that which is despised and daunting. Could fear of the feminine be not unrelated to the removal of infinity from Mathematics at the moment of the latter's birth? Could the parallel courses of these threads explain the apparent urge to explain infinity clearly and finally, placing it within manageable bounds and therefore under Mathematical jurisdiction? Is the restriction of the infinite to Mathematical laws an attempt to restrict, dominate, and control the deeply uncontrollable concepts of infinity, physically represented by the eternal cycle of birth and death, by unimaginable reaches of time and space? Is it ludicrous to think of the program to domesticate the infinite (and, failing to do so, exile it) as analogous to man's project of taming woman's sexuality, so often depicted in such terms as "voracious" and "unstoppable"—and "infinite"?

Infinity Goes to School

> Maggie [age four] told me in the Wal-Mart parking lot yesterday, as we were climbing on mountains of earth and rocks plowed up to destroy even more formerly holy Native American ground, that infinity is 666 with a bunch of 6s added on to it until you couldn't add any more 6s at all.
> Our friend Michael Rose

> About infinite decimals....Of course I know about them, but that's not what they're teaching in school (if they actually TEACH at all). Algebra is what we're supposed to be learning. I think infinite decimals are cool, but just because they're infinite. I love the theory of infinity. That's even cooler.
> Bill's 10-year-old cousin Kayla

Perhaps the impossibility of a four- or ten-year-old's being willing or able to appreciate a formalization of the infinite explains why it has been expelled from school mathematics and science. Maybe, as we spoke about at the start of this chapter, the unavoidable paradoxes glued to the infinite account for the paradox that the only mathematical topic for which your average citizen has any passion is the mathematical topic most unwelcome on school grounds. And/or it might be that infinity's excused absence is due to its association with the masculine Judeo-Christian-Muslim God running up against the firewall separating church and state.

A few brave and wise souls have dedicated themselves to bringing the infinite into school mathematics. We direct your attention to three instances: Mills, O'Keefe, & Whitin (1996); Cordeiro (1990); and the finest children's story about mathematics ever written, Nancy Casey's 1993 yarn "The Hotel Infinity," which has regaled and edified eight-year-old children, preservice elementary- and secondary-mathematics teachers, and doctoral students in a class on mathematical epistemology. In addition, Bill has had a modicum of success engaging and enraging students with the question of whether or not the infinite decimal expansion .999...is equal to 1 (Rosenthal, 1999).

We have been unable to locate any school-science curricular materials pitched toward the infinite. Opportunities nevertheless abound. Infinity has to be actively suppressed in order not to arise in discussions of motion, time, cosmology, and the divisibility of matter, among other topics. Listening to students will inevitably yield a plethora of questions involving the infinite, such as the one Elaine recently heard in an elementary-science-methods class studying genetics: "If children are a combination of parents, can there be an infinite number of combinations?" Teachers who read widely and well will encounter concepts and contexts inhabited by infinity. Claude Allegre (1992) offers one possibility over which we recently stumbled.

> Werner's Neptunian theory and his six stages of geologic development must be placed in this chronological context [that of the Bible]. We know that Hutton, Playfair, and then Lyell, however, utterly opposed the idea of time as closed and *finite*. Their theory of Uniformitarianism gave geologic time an *infinite* duration. Just as the motion of the planets seems to have neither a beginning nor an end, geological phenomena repeat themselves over an infinitely long period of time. These phenomena are permanent fixtures of nature. Infinite time is an extremely useful hypothesis in geology because it completely eliminates the idea of an exact age of the Earth and therefore the need for the absolute measurement of geologic time. Since phenomena repeat themselves over and over, it is neither necessary nor important to establish a chronology which, in any case, would be lost in infinity. Although Hutton was correct in his analysis of geologic causality, his views long warped our understanding of time. (pp.38–39)

This piece of a larger historical debate illustrates how people have used the infinite to model physical phenomena (here, the age of the Earth) in constructing scientific knowledge.

The patently gendered history of the infinite further recommends it as a topic to be treated in schools. Our small history provides seed corn for speculating on and imagining the curricular power of the infinite. The role of infinity in Mathematics and science is controversial and uncertain. School curricula and instruction regularly avoid the controversial and uncertain. This avoidance is pandemic in Mathematics and science, not the least because controversy and uncertainty in these areas challenge the myth of absolute foundations for their knowledge. For this very reasoning, we suggest that teachers, curricularists, administrators, and teacher educators welcome infinity into the classroom.

> 1. A study of the place of infinity in the ongoing development of mathematical thought will provide students with a more accurate vision of the *social* construction of math, with all its fascinating human complications.
> 2. Because infinity is found in all human expressions of experience, a study of infinity allows for interdisciplinary explorations that (i) are often recommended by multicultural educators (e.g., Atwater, 1996); and (ii) have the potential to undo the mathematical alienation disproportionately suffered by girls and women, persons of color, and children in poverty.
> 3. People are intrigued, disturbed, and inspired by the infinite. It is unethical to deprive children of the opportunity to participate in this age-old and worldwide

discussion.

4. Whatever the reasons that neither we nor Elaine's 23-year-old daughter nor Bill's cousin Kayla had the chance to take up infinity in school, we want matters to be different by the time our friend Maggie makes it to, say, fourth grade. We had originally planned a more extensive argument for the inclusion of infinity in school mathematics and science. Now, however, we believe that a lengthy case-pleading is unnecessary, even gratuitous. Let's just put a few pieces together. A venerated Mathematician asserts that infinity is what mathematics is essentially about. ("Mathematics is the science of the infinite.") Even those of us who don't want science epistemologically or pedagogically tethered to mathematics want students to have opportunities to learn math and science to feed and nurture each other. There's no better topic for interdisciplinary study. Elaine is still mad that she was curricularly deprived of infinity. Kayla thinks infinity is cool. Maggie is gracing adults with her theories of the infinite before she reaches the age of five. Do we *really* need to elaborate further?

Acknowledgment

We are deeply indebted to Whitney Johnson of Michigan State University for the inspiration, insight, and information she has provided to us.

Endnotes

1. In addition to works cited elsewhere in this chapter, see: Bley (1998); Falk (1994); Lavine (1995); Holt (1999); Maor (1987); Moore (1990); Pickover (1995); and Vilenkin (1995).

2. We have investigated our own infinity-related interpretations and associations in other work (Howes & Rosenthal, 1995). Non-Western infinities worthy of far more mention than they have received include those of the Jaina civilization of the Indian subcontinent (ca. 500 B.C.E). (Joseph, 1991/1992) and the Iqwaye of Papua, New Guinea (Mimica, 1988).

3. There has been other good recent work, some of it flagrantly feminist, on how males created Mathematics as a fraternal organization. Mary Harris (1997) testifies: "The same patriarchs, the same philosophies, the same preaching and the same practices that confirmed mathematics as the highest form of human intellect, placed woman as the lowest form of human life" (p. 27). See also Suzanne Damarin (1995).

References

Allegre, C. (1992). *From stone to star: A view of modern geology.* Cambridge, MA: Harvard University.

Atwater, M. M. (1996). Social constructivism: Infusion into the multicultural science education research agenda. *Journal of Research in Science Teaching 33*(8), 821–837.

Bishop, A. J. (1988). *Mathematical enculturation: A cultural perspective on mathematics education.* Boston: Kluwer Academic Publications.

Bley, M. (1998). *Reasoning with the infinite.* Chicago: University of Chicago.

Casey, N. (1993). The hotel infinity. In Nancy Casey & Michael Fellows. *This is mega-mathematics!: Stories and activities for mathematical thinking, problem-solving and communication*: Los Alamos National Laboratories, Los Alamos, New Mexico. (Available at http://www.cs.uidaho.edu/~casey931/mega-math/welcome.html)

Cordeiro, P. (1990). Playing with infinity in the sixth grade. *Language Arts 67*, 557–566.

Damarin, S. K. (1995). Gender and mathematics from a feminist standpoint. In Walter G. Secada, Elizabeth Fennema, & Lisa Byrd Adajian (Eds.), *New directions for equity in mathematics education* (pp. 242–257). New York: Cambridge University.

d'Aragona, T. (1997). *Dialogue on the infinity of love.* Edited and translated by Rinaldina Russell and Bruce Merry. Chicago: The University of Chicago Press (Original work published 1547).

Dauben, J. W. (1979). *Georg Cantor: His mathematics and philosophy of the infinite.* Cambridge, MA: Harvard University.

Dauben, J. W. (1995, June). Cantor and the epistemology of set theory. Paper presented at the Annual Meeting of the Canadian Society for History and Philosophy of Mathematics, Montréal.

Dinnerstein, D. (1991). *The mermaid and the minotaur: Sexual arrangements and human malaise.* New York: HarperPerennial. (Original work published 1976)

Eisler, R. (1987). *The chalice and the blade: Our history, our future.* New York: HarperCollins.

Falk, R. (1994). Infinity: A cognitive challenge. *Theory & Psychology 4*(1), 35–60.

Griffin, S. (1978). *Woman and nature: The roaring inside her.* New York: Harper & Row.

Harris, M. (1997). *Common threads: Women, mathematics and work.* Staffordshire, England: Trentham Books Ltd.

Holt, J. (1999, 20 May). Infinitesimally yours. *New York Review of Books*, 63–67.

Howes, E. V., & Rosenthal, B. (1995). Infinity (continued): What's gender got to do with it? Paper presented at the Seventeenth *JCT* Conference on Curriculum Theory and Classroom Practice,

Monteagle, Tennessee.
Howes, E. V., & Rosenthal, B. (1996). Less than Zeno: What *they* leave out of the infinite. Paper presented at the annual meeting of the Canadian Society for History and Philosophy of Mathematics, St. Catharines, Ontario.
Johnson, W., & Rosenthal, B. (1996). The reflection of early Greek mathematics in the mathematics of today. Paper presented at the Annual Meeting of the Canadian Society for History and Philosophy of Mathematics, St. Catharines, Ontario.
Joseph, G. G. (1992). *The crest of the peacock: Non-European roots of mathematics*. New York: Penguin. (Original work published 1991)
Kirk, G.S., & Raven, J.E. (1957). *The Presocratic philosophers: A critical history with a selection of texts*. Cambridge: Cambridge University.
Lakatos, I. (1976). *Proofs and refutations: The logic of mathematical discovery*. Cambridge: Cambridge University.
Lavine, S. (1994). *Understanding the infinite*. Cambridge, MA: Harvard University.
Lear, J. (1980). Aristotelian infinity. In A.W. Moore (Ed.), *Infinity* (pp.55–78). Aldershot, UK: Dartmouth, 1993.
Lerner, G. (1986). *The creation of patriarchy*. New York: Oxford University.
Lerner, G. (1993). *The creation of feminist consciousness*. New York: Oxford University.
Maor, E. (1987). *To infinity and beyond: A cultural history of the infinite*. Boston: Birkhäuser.
Maziarz, E. A., & Greenwood, T. (1968). *Greek mathematical philosophy*. New York: Frederick Ungar.
Mills, H., O'Keefe, T., & Whitin, D. J. (1996). *Mathematics in the making: Authoring ideas in primary classrooms*. Portsmouth, NH: Heinemann.
Mimica, J. (1988). *Intimations of infinity: The mythopoeia of the Iqwaye counting system and number*. Oxford: Berg.
Moore, A.W. (1990). *The infinite*. New York: Routledge.
Moore, A.W. (1995, April). A brief history of infinity. *Scientific American*, 112–116.
Neumann, E. (1994). *"The fear of the feminine" and other essays on feminine psychology*. Princeton, NJ: Princeton University.
Noble, D. F. (1992). *A world without women: The Christian clerical culture of Western science*. New York: Oxford University.

Pickover, C. A. (1995). *Keys to infinity*. New York: John Wiley & Sons.
Ring, M. (1987). *Beginning with the Pre-Socratics*. Mountain View, CA: Mayfield Publishing Company.
Rosenthal, B. (1999, April). Why .999...is AND is not equal to 1 (A social-scientific proof). Paper presented at the Annual Meeting of the American Educational Research Association, Montréal.
Rotman, B. (1993). *Ad infinitum: The ghost in Turing's machine*. Stanford, CA: Stanford University.
Rucker, R. (1982). *Infinity and the mind: The science and philosophy of the infinite*. Boston: Birkhäuser.
Schiebinger, L. (1989).*The mind has no sex?: Women in the origins of modern science*. Cambridge: Harvard University.
Schimmel, A. (1993). *The mystery of numbers*. New York: Oxford University Press.
Stone, M. (1976). *When God was a woman*. San Diego: Harcourt Brace & Company.
Sweeney, S. J. (1972). *Infinity in the Presocratics: A bibliographical and philosophical study*. The Hague: Martinus Nijhoff.
Thomas, R. (1996). Proto-mathematics and/or real mathematics. *For the Learning of Mathematics 16*(2), 11–18.
Vilenkin, N. (1995). *In search of infinity*. Translated by Abe Shenitzer. Boston: Birkhäuser.
Wertheim, M. (1995). *Pythagoras' trousers: God, physics, and the gender wars*. New York: Times Books.

Chapter Eight
Cookbook Classrooms; Cognitive Capitulation
Dave Pushkin

> I know myself, and my tendency to be lazy...I like taking the path of least resistance, and I could see myself doing that as a teacher, especially if I want tenure. You can't make waves for a few years, and you have to do what everyone else does, even if you disagree with it. It scares me...I can see myself becoming one of them, simply because fighting it can give you an ulcer. It just isn't worth it...I'd rather do something else and have peace of mind.

These are the words of Jim (a pseudonym), a once-aspiring high school physics teacher, and a former general chemistry student of mine. Several months ago, Jim informed me that he was changing his major from physics to computer science, and was dropping out of the teacher education program. When I asked why, he reflected on his experiences as a science student, a student in teacher education courses, and an observer of classroom practice, both in high schools and his university. Jim was convinced the "good" pedagogical things (e.g., constructivism) he heard about in education courses, and even witnessed in my general chemistry course, were aberrations, exceptions to the norm. After all, if these "good" things were really "good," why weren't more faculty exposing students to them (Pushkin, 1999)? Jim had other concerns regarding his future context as an educator and potential agent of cultural change, but the perceived discontinuity between what he heard and saw ultimately disturbed him enough to give up the idea of a teaching career.

Yet, as we look deeper into Jim's words, we see profound notions for a twenty-year-old:

- path of least resistance
- want tenure
- do what everyone else does
- becoming one of them
- scares me
- just isn't worth it

What we observe are the subtle symptoms of *cognitive capitulation* (Pushkin, 1998a), where gradually over time, intelligent educated individuals no longer think for themselves, but for their hierarchical context. It is essentially a surrendering of one's thinking process; arbiters of a hierarchy think for them. If Jim's thoughts only stopped at the third notion, he would fit in beautifully within his hierarchy. However, because Jim thought beyond that point, whether as resistance or fear, he realized the potential danger of remaining within a cultural context that disturbed him.

Cognitive capitulation is a process by which learners (and educators) blindly surrender their thinking processes to arbiters of knowledge and power. In the hegemonic, inertial, and hierarchical culture of traditionalist science, too many learners are still indoctrinated to be obedient passive recipients of fragmented knowledge. Success is still measured in terms of reiterating quantity upon command, following unquestioned standardized directions, and demonstrating competence to validate instruction (Kincheloe, 1991 and 1993; Pushkin, 1998b). Teachers of science are still expected to stay within the lines and deliver a nauseatingly familiar, linear progression of disjointed and decontextualized content. Traditionalist classroom science instruction is routine, predictable, and contrived, essentially straight from a pedagogical cookbook.

A Theoretical Framework

In 1970, Harvard psychologist William Perry presented four stages of adult cognition (Kincheloe, 1993; Perry, 1970; Pushkin, 1997a): Dualism, Multiplicity, Skepticism, and Relativism. Although educators may question the validity and generalizability of Perry's research, due to the demographic homogeneity of his student sample population, similar characteristics of these cognitive stages can be observed among both genders and multiple ethnicities (Pushkin & Colón-González, 1998).

Perry's first stage, dualism, reflects a dichotomous viewpoint. Dualistic thinkers see the world in terms of black/white, right/wrong, and true/false. Unable to recognize or respect alternative views, dualism is very analogous to solipsism (e.g., Staver, 1998).

For example, consider a chemist's definition of "diffusion," or a physics student's definition of "entropy" (e.g., Pushkin, 1995). The chemist views diffusion primarily (if not always) in terms of mixing gases at thermal equilibrium. Is this how novices look at diffusion, or for that matter, other branches of science (e.g., biology and physics)? It would be doubtful. Simply ask a teenager why s/he sleeps with a textbook under her/his pillow the night before an exam. The cliché *learning by osmosis* is inadvertently a *pseudoconception* (Pushkin, 1995, 1996, & 1997b) of diffusion; however, it was associated with a sense of flow or movement.

Biologists and physicists both look at diffusion in terms of particle flow along a concentration gradient, whether in the context of dialysis or dielectrics. Even biochemistry textbooks address diffusion similarly. But a general chemistry instructor (possibly a physical or analytical chemist by trade) recognizes only the view of her/his own course textbook. If the so-and-so book of general chemistry states diffusion in terms of mixing at thermal equilibrium, by golly that's the correct definition. All other definitions are wrong and not even equivalent. We're not teaching biology or physics; we're teaching chemistry, and we view things the chemistry way!

The multiplist brings her/his own view to a learning environment yet soon realizes the view of that environment's authority figure carries significant weight. For example, consider a student who believes a certain vegetable oil is denser than water because of its "thicker" appearance. A science instructor may offer multiple evidences to contradict this mis/pseudoconception:

- Oil floats on top of water
- Equal volumes have different masses
- Equal masses occupy different volumes
- Comparisons of boiling points with kinetic molecular theory implications
- Comparisons and distinctions between density and viscosity

Perhaps the student will see enough value in this evidence to conceptualize the oil as less dense than water. Perhaps s/he will simply agree with the notion of oil being less dense than water on the basis of the instructor's evidence outnumbering her/his own conception. *If the majority rules, the majority must be correct—when in Rome, do as the Romans.* This cliché indicates deference to a view other than one's own.

However, the question remains as to whether or not the student internalizes this conception. S/he may indeed accept that oil is less dense than water and even answer relevant test items correctly. Yet the possibility exists of students leaving science courses conceptualizing density as a function of thickness rather than relative compactness of particles. Despite deference and correct test answers, students may still maintain their own original conceptions. The multiplist potentially thinks one view yet answers according to a more-dominant view.

In Perry's third stage, skepticism, students develop a questioning mechanism toward knowledge and its use. They view knowledge as more than "right/wrong" answers; they view knowledge as to whether or not it makes sense.

> Today's lab was just so great...apparently students are too stupid to follow the lab handout, so we were given a simplified supplemental handout. The lab was cut into one period instead of two...and we were actually done in less than an hour.
>
> My lab partner tried to explain to me why this was good, and how the original handout was too complex...the how's and the why's aren't important...only the what is. I don't just want to understand the what; I want to also understand the how's and why's.

Cathy, an aspiring chemical educator, partially described her unfulfilling experiences a year ago in a second semester general chemistry course (Pushkin, 1999). She sought more from knowledge beyond her professor's lecture notes yet questioned its value or purpose within a hegemonic, depowering, and stupidifying context. Very little changed within a year's time; Cathy eventually expressed frustration with her organic chemistry course.

> Although I am learning more in [Dr. A's] organic class this semester as opposed to [Dr. B's] last semester, my desire to learn has been stifled and nearly destroyed. [Dr. A] has no tolerance for questions even if they are thought out and intelligent. You simply have to learn what he says to learn ... question NOTHING! If I didn't realize it before, I now know that independent thought is at best unwelcome at [her university].

Although Cathy's personal bias toward captivity and traditionalist lectures is quite clear, she does start to consider some pros and cons on her learning experiences: On *one hand, this professor's a jerk, but on the other hand, I learn more from him than Dr. Congeniality.*

The fourth of Perry's stages, relativism, represents the broadest and most empathetic level of thought. The ultimate relativist is one who can step out of her/his own shoes and into someone else's.

Why? In order for the relativist to take a position on an alternative view, s/he must be able to observe and reflect from the alternative vantage point, evaluate for potential strengths and weaknesses, then decide whether to accept or reject this view. The key attribute of

Cookbook Classrooms 197

relativism is perhaps the ability to see things from others' vantagepoints. While Cathy viewed her recent chemistry professors' pedagogical practices as insulting, dehumanizing, and demoralizing, when viewed from the professors' perspectives, however insensitive and unnurturing they may be, Cathy's insights may not be broad enough in terms of seeing all sides to the situation. On the other hand, Ruth, an aspiring biology teacher, attempted to see her general chemistry professor's perspective by individually discussing her frustrations with him.

> My professor freaked out the other day...he asked the class a question; so I answered it. He looked stunned, as if I did something wrong. Apparently he only asks rhetorical questions and answers them himself. I broke a rule of interrupting him. In fact, he announced on the first day that we are to ask him questions before class starts, after it ends, or during office hours...so I guess I wasn't supposed to do that. He actually had trouble regrouping himself...it took him a few minutes to get back into his mode of just telling us stuff.

> It frustrates me, because all he does is repeat the book...only the book, and nothing beyond it. I asked him during office hours why he has to spoon-feed us and bore me, and he said "what do you want me to do...fail everyone in the class? I can't teach any other way, or else the unfortunate majority won't get it and fail."

Is Perry's progression of cognition simply an independent, internalized mechanism for learners? In our hegemonic culture of science, most likely not. If one looks at learning from a Vygotskian perspective (e.g., Driscoll, 1994; Moll, 1990; Pushkin, 1997a), three key components of Vygotsky's theory play into the context:

- Zone of Proximal Development,
- Role of Teacher as Mediator, and
- Impact of Social Discourse.

The Zone of Proximal Development is essentially an individual's learning curve, which translates to learning having its own pace. Some students "get" concepts right away in a science class; some struggle until eventual understanding is reached. Metaphorically, in a standard science lecture course, the "ugly duckling" from the midterm may indeed become the "beautiful swan" by the final exam. This feature of Vygotsky's theory practically tells science instructors not to direct struggling students to drop a course early on; they may actually enjoy ultimate success by sticking with the course.

According to Vygotsky, the classroom teacher, an authority figure of knowledge, can lead students toward understanding and success, by scaffolding the direction of thinking. Unlike Piagetian theory, which often places the progression of thinking and learning solely within the hands of a student, Vygotskian theory places enormous responsibility on teachers to carefully lay out proverbial "breadcrumbs" for students to follow toward a desired learning goal. Learning is neither independent of teachers nor overtly dictated by teachers; teachers *guide* it.

The same can be said about social discourse. As White and Gunstone (1992) surmised, "Learning does not take place in a vacuum." Students do not simply learn within themselves; external socio-cognito-cultural influences shape learning as well. Not only are students exposed to teachers but to fellow classmates, who may have their own conceptions or perspectives about scientific knowledge. Look at how cooperative groups function, or lab partners/teams. Which view eventually becomes the group's view? The most dominant (correct or not). How does this come about? Dialog.

However, the most subtle and profound meaning we can derive from Vygotsky's theory is the potentially powerful role that teachers, or authority figures, have in the cognitive development of others (students or other teachers). A colleague once told me: Teaching is like a conversation, and some learning should take place as a result of that conversation. There really is no such thing as "bad" teaching; teaching can be "good," or it can be inconsequential (Pushkin, 1995). Is this to say teaching is either good or at worst benign? Is this to say teaching cannot stifle and destroy a desire to learn, as Cathy shared, or frustrate, as Ruth shared? What about Jim's perceived need to abandon dreams of a teaching career for self-preservation? Teachers, and teaching practices, have enormous power to stimulate, enhance, reinforce, or inhibit thinking, learning, or alternative pedagogies. "Good" teaching can be a very positive external influence on students; "bad" teaching can be very negative.

However, the "benign" teaching we see in traditional science lecture courses can be the most negative influence of all. Traditionalist lecturers consider themselves "good" if not "excellent" teachers or instructors. What is often a platitude for Distinguished Professor or Professor/Teacher of the Year? S/he is a wonderful lecturer, sharing the wealth of her/his expertise with eager students. In essence, these are scholars who can "deliver the goods," covering the content with crisp, calculated, effective precision. Each lecture hour offers a proportional amount of material to validate the course syllabus and exam schedule. Some students survive the ordeal; many do not and change majors.

We speak of learning objectives and testing expectations—sometimes very traditional and content oriented, sometimes global and general—but does the class culture reflect this (Hammer, 1995)? Together, teaching and learning are components of an ongoing dialog; what we teach needs to build on what students know, not what the syllabus states. It is detrimental to teach in a manner that leaves students "in the dust"; despite the university structure, the definition of pedagogy is not Darwinism.

Cathy's recent organic chemistry professor, Dr. A, was awarded Distinguished Professor and Professor/Teacher of the Year honors during his career; he is considered a wonderful lecturer. He comes to lecture hall ready for business. He lectures and delivers his goods to passive notetakers and is rarely derailed from his appointed duty. Why? No tolerance for questions, as Cathy shared. In fact, Cathy's second-semester general chemistry professor, although not as honored as Dr. A, had virtually no tolerance for questions either. Consider another shared insight from her general chemistry ordeal:

> It's sad...he gave a test on kinetics, and even though I got the problem wrong, so did he in his solutions. I went over the data he gave us, and the reaction is first-order, not the second-order he said. When I tried to bring it to his attention, he didn't want to hear it, and chewed me out for questioning his authority and showing disrespect. Talk about humiliating...talk about dehumanizing. You can't ask questions in class...I raised my hand, and he yelled at me to put my hand down and stop interrupting him. He was writing down the wrong reaction, and everyone just sat there and copied down exactly what he wrote like obedient little stooges.

I will address this particular insight in more detail later, but it's apparently clear how professors' perceived need to "stay on track or schedule" plays into their teaching. Their syllabi are bloated to unreasonable limits with voluminous textbook content; every concept or topic is a sacred cow that cannot be eliminated (e.g., Pushkin, 1995). So much material in such little time. Some "good lecturers" actually construct their syllabi according to the following logic:

> A semester consists of 15 weeks, where we meet hourly 3 times per week. We need to cover 10 chapters from the textbook, which involves approximately 500 pages of material (not counting figures and end of chapter questions). If we eliminate 5 meetings for tests or holiday, that leaves approximately 40 hours of instruction. 500 divided by 40 equals 12.5 pages of material, so I need

to cover 12.5 pages of material every hour. That makes sense...this translates to approximately 4 lecture hours per chapter.

To be interrupted by a student during lecture time is potentially catastrophic as Ruth observed of her professor (a recently tenured professor and an "excellent teacher"). To interrupt (or disrupt) with questions creates a tense situation for lecturers; do they keep priorities "straight," or do they "indulge" students and fall behind schedule? What is a "good" teacher to do, assuming they understand what constitutes "good" teaching?

Connecting Theory and the Story

The method of *Currere*, a writing style forever linked to Bill Pinar, personalizes one's research. Sharing one's perspective of self, a researcher not only presents growth within an area of knowledge but growth within those who study it. Pinar (1994) defines *Currere* as "regressive-progressive-analytical-synthetical" (p.19). Regressive, because it looks to the foundation of the research. Progressive, because it looks to the implication of the research. Analytical, because it looks to the conceptualization of the research. Synthetical, because it looks to the connection of the research. As you are reading this chapter, you notice a phenomenon (cognitive capitulation), personal stories from aspiring and practicing teachers, and a grounded theory, all interwoven. In the final section of this chapter, you will experience various manifestations of cognitive capitulation in science classrooms at both the secondary and university levels.

Why tell stories? It is a form of presenting qualitative research, where a narrative approach can present a moral (e.g., Coffey & Atkinson, 1996). Ethnographic methodologies allow us to examine the human condition (e.g., Grills, 1998; Hammersley, 1992; Scheurich, 1997; Seale, 1998). Storytelling is an extension of conversational analysis (e.g., ten Have, 1999); stories can even be analyzed as data (e.g., Ollerenshaw, 1999). Charles Dickens wrote books of social satire to help his generation understand the human condition, to teach people something about life, social dynamics, and discourse. The stories we tell enable us to understand phenomena and theory/practice issues through the experiences of others (e.g., Valdés, 1998).

Currere need not center on the personal experiences of a researcher to shape scholarly discourse. In this chapter, the regressive aspect comes from the identification of cognitive capitulation. The progressive aspect will come from the manifestations of cognitive capitulation. The analytical aspect comes from exploration of Perry's and Vygotsky's cognitive theories. The synthetical aspect will come as I describe how cognitive capitulation occurs and how we observe it among students and teachers. The stories we tell eventually lead to an interpretation. The interpretation eventually leads to a theory that certain stories validate.

Manifestations of Cognitive Capitulation

> During the phase of the closed society, the people are *submerged* in reality. As that society breaks open, they *emerge*. No longer *mere spectators*, they uncross their arms, renounce expectancy, and demand intervention. No longer satisfied to watch, they want to participate. This participation disturbs the privileged elite, who band together in self-defense. (Freire, 1973, pp.13–14)

How does cognitive capitulation come about? Because traditionalism "knows" what is best for students, artificial criteria are established for learning based on presumptions of incapability and ignorance. Traditionalist teaching, according to these presumptions, limits the depth and breadth of students' learning experiences. Students are led to believe they've learned an extensive amount of content, when in reality, they've mastered a superficial amount of factoids (e.g., Kincheloe, 1991 and 1993; Pushkin, 1998b). Traditionalist testing, according to these presumptions, evaluates to demonstrate ignorance versus comprehension, validating a paradigmatic fulfilling prophecy.

Students are set up for a "fall"; their poor performances force them to rely on their arbiters even more, as if they cannot learn without them. With success and rewards the ultimate goals, students (willingly or reluctantly) place themselves into the hands of arbiters, expecting to be *shown the way*. Arbiters, unfortunately, do not show the way toward success; they show the way toward perpetual helplessness and cognitive stunting, much along the lines of what Macedo (1994) calls *stupidification*. Even if students achieve academic success via submission to arbiters, they will never understand or appreciate science's wonderful depth and breadth. Never do the following clichés apply more:

> All the book smarts in the world and no sense to come out of the rain. A 4.0 average only means lots of right answers on tests; it doesn't mean you know anything.

Consider the following exchange of classroom experiences by Ruth, Cathy, and Jim during their second semester general chemistry course, each having a different instructor:

Ruth:
I'll tell you guys something...the atmosphere is really different from last semester. The classroom seems to cater to the people who just want everything spoon-fed to them...Professors need to have a little faith in their students.

Cathy:
We spent the first two classes] going over his final from last semester, and I had to strain to keep from laughing. It was such a joke...most of the questions were so idiotically simple I can't believe students did as poorly as they said. My professor just dictates everything in the simplest form for idiots to understand, but he tests us on stuff that's never covered in class. Lucky for me I understand the material...everyone else is hopelessly lost and can't solve his problems.

Jim:
The lecture consists basically of my professor writing all of the definitions of the chapter on the board, showing us the transparencies that are photocopied right out of the book, and showing us the same examples that are given in the book. I feel like all the knowledge I learned [last semester] is slowly seeping out of my head because I'm really not using it...I've been lulled into stupidity. Her tests are a joke...the first part of the test is vocabulary terms for definition, then problems like the chapter. And most of the kids still can't do them! I myself do not necessarily understand how I'm getting the answer, since I'm blindly following her recipe for solving the problems.

Cathy:
I think they teach us this way because they really think we're stupid and can't learn. I went to office hours for help on something (clarification of his gibberish mostly) and asked a question about applying the concept to other reactions. He just stared at me and told me it's inappropriate to ask about things that are irrelevant to what he teaches us...as if we're only supposed to know what he wants us to know.

Last semester we were expected to do this and any "applied" questions that we would have were more than welcomed to be asked. This semester, "understanding" is bad and only "troublemakers" attempt to do it.

Ruth:
I think there's some truth to these professors teaching us to keep us down. If we only do what they tell us, then we'll never get wise enough to challenge them.

Jim:
That's exactly the way my professor does it. Everyone in the class is hopelessly dependent on her for everything, including their grade. She's easy enough, that if you have some fraction of a brain, you can get a 'B' and not learn anything anyway.

Ruth:

My professor's problems are so easy, you have to have no brains to get them wrong. It's as if he needs to know everyone can at least give him back the answers he wants. I guess that's just the way it is...why bother fighting it?
Cathy:
Let's face it...we're supposed to be stupid little children who need to learn obedience. That's what this semester's been all about. It's not about learning.

Notice Cathy's defiance toward her classroom context. Earlier in this chapter, I shared her disillusionment, not only from lack of intellectual challenges but also from the stifling suppression of her professors during lectures. Cathy was told to keep quiet, do as told, not interrupt, and learn respect.

Sheila Tobias considers this a means to develop students' OQ (Obedience Quotient) as opposed to their IQ (e.g., Tobias & Tomizuka, 1992). This is also described as an *acculturation* and indoctrination to a cognitive apprenticeship (Duit & Treagust, 1998; Gilbert & Boulter, 1998). Rather than empowered, students are rendered *depowered*, traumatically suppressed by their invisible subordination to an ideological manipulation, as they become pedagogically automated robotic technicians (e.g., Bartolome, 1998; McLaren, 1998; Wong, 1998). Cathy fights this process with all her internal strength and self-identity.

Cathy's traumatic suppression existed in many forms. When she was "chewed out" for interrupting and questioning her general chemistry professor, he actually asked her if she had a neuromuscular problem with her arm. When she visited him during office hours, more than once, she left in tears; eventually she gave up on office hours. She also came to realize mere obedience would not translate to empowerment, understanding, and independence as a critical thinker. Obedience would only lead to more dependence and less understanding, as she noted among her classmates as "stooges." Ultimately, these "stooges," or variations of them, would not be capable of thinking for themselves as Cathy, Ruth, and Jim all surmised.

> Assistencialism is an especially pernicious method of trying to vitiate popular participation in the historical process. In the first place, it contradicts [our] natural vocation as Subject in that it treats the recipient as a passive object, incapable of participating in the process of [one's] own recuperation; in the second place, it contradicts the process of "fundamental democratization." The greatest danger of assistencialism is the violence of its anti-dialogue, which by imposing silences and passivity denies [them] conditions likely to develop or to "open" their consciousness. For without an increasingly critical consciousness [they] are not able to integrate themselves into a transitional society,

marked by intense change and contradictions. Assistencialism is thus both an effect and a cause of massification.

> The important thing is to help [them] help themselves, to place them in consciously critical confrontation with their problems, to make them the agents of their own recuperation. In contrast, assistencialism robs [them] of a fundamental human necessity—responsibility...Responsibility cannot be acquired intellectually but only through experience. Assistencialism offers no responsibility, no opportunity to make decisions, but only gestures and attitudes which encourage passivity...this method cannot lead...to a democratic destination. (Freire, 1973, pp.15–16)

So what is cognitive capitulation in theoretical terms? If we look back to Perry's first two stages of cognition, dualism and multiplicity, we observe a progression in which students evolve from dichotomous thinking to a deferent dichotomy. We need to remember, though, the multiplist still maintains her/his own conceptions when not in the presence of an arbiter or authority figure.

However, there appears to be a point during an undergraduate science education where adults no longer hold onto their own conceptions. Their deference is no longer strategic and contrived; it becomes their knowledge base. The longer students remain in the hegemonic culture of traditionalist science, the closer they get to the hierarchy of power. To become a master, an apprentice needs to think like a master; what the master knows becomes what the apprentice knows. Essentially, the apprentice identically defines what is "right" as defined by the master; the same occurs for "wrong." The truth known and dispensed by the master becomes the only truth the apprentice knows.

That said, what we ultimately observe is not a progression of thinking; we observe a *regression*, a regression from multiplicity to dualism. Students develop multiplistic thinking skills early in their undergraduate science courses. To "play the game" and achieve academic success, students' *modus operandi* becomes *Tell us what you want to hear, and we'll tell you what we think* (Perry, 1970; Pushkin, 1995). As time goes on, students' *modus operandi* becomes *Tell us the truth and we'll agree to it and preach it*. How do general chemistry professors define diffusion? How do general physics students define entropy? Exactly verbatim according to their arbiter, the textbook. Truth is only what we dispense and, therefore, know.

Consider the touching words that Cathy, Ruth, Jim, and other students inscribed to me on a plaque and card of appreciation for their general chemistry experience with me:

> Your aim as our educator was to teach us how to think rather than what to think. Thank you for this, for unsettling our thoughts, awakening our curiosity, and kindling our minds. In your classroom, education was always achieved rather than received. For inspiring us to be life-long learners, we are forever indebted to you. You are a facilitator of knowledge rather than an authority seeking to shove ambiguous concepts down students' throats. You are aware that a classroom containing massive amounts of information is not equivalent to a massively informed classroom. We aspire to influence our students in the manner in which you affect so many of yours.

This is why teachers are potentially such a powerful external socio-cognito-cultural influence on students. Teachers can either enhance the thinking and learning process or completely inhibit it. Teachers can either nurture independent critical thinking or completely stunt it. Teachers can either emancipate empowered learners or enslave them as helpless depowered robots. Teachers can either guide a new generation of intellectuals or program a perpetuation of obedient capitulators. Obedient capitulators inevitably become the future teachers of traditionalist science. Inertial pedagogy and curricula help perpetuate a species of dualists; cognitive capitulation helps preserve a paradigm.

However, this capitulation is not limited to science students. Teachers of science are expected to maintain the sanctity of the paradigm; arbiters of knowledge are essentially Darwinist gatekeepers to a power structure. For students to remain at the knowledge level of apprentice, teachers must be careful to limit students' access to knowledge. Teachers failing to strictly enforce that limit threaten the chain of command and balance of power. As conduits to power, such teachers are viewed as heretical traitors; they are quickly eradicated from the hierarchy as inefficient and incompetent. Only the fittest students honor the chain of command; only the fittest teachers honor the balance of power.

Junior science faculty, for academic self-preservation, must submit themselves graciously to the leadership of departmental elders. Because introductory science courses are so contrived and orchestrated, teachers need to follow a script, a pedagogical cookbook predetermined by an arbitrary philosophy. Assessment is as much a measure of pedagogical consistency as of cognitive stunting. Junior faculty are the ultimate multiplists, adopting the arbitrary philosophy until it becomes internalized and naturalized. The ultimate reward for pedagogical conformity, the commission for their consignment, is tenure and promotion. Cognitively backwards, the mature and successful science faculty member is a dualist. Senior faculty learned to narrow their views as they honed their skills. Senior faculty are the embodiment of successful capitulation; they possess the greatest OQ and are ultimately rewarded for it.

> Mike:
> It's funny...when I first joined the faculty, the dean asked me to submit my personal philosophy of teaching and learning. I struggled for hours, sitting in front of a blank computer screen, trying to figure out what my philosophy was and how to put it in writing. I thought I had a vision of what my lecture classes should look like, but writing about it wasn't easy. One of the senior faculty [Dr. A] came by my office, and handed me a two-page thing, and said "You don't need to worry about that stuff; here's your philosophy of teaching and learning." It was basically the same thing everyone used when they first got here. I thought it was weird, but these are the guys who will determine my tenure. Hey, five years isn't so long... eventually I'll get my tenure and can develop my own way of teaching general chemistry. The only problem is it's five years later, and I can't remember what I wanted to do anymore. Oh well; at least I have a routine down pat.

Mike is an Ivy-League trained chemist entering his seventh year as a professor at a teaching university, tenured but not yet promoted. Mike was Ruth's second semester general chemistry instructor and serves as the departmental certification officer for prospective high school chemistry teachers. Mike claims to care about teaching, yet Ruth perceived something different. From Ruth's assessment of Mike's teaching, he is the ultimate *depowerer* of students, the perpetuator of cognitive capitulation. His ultimate curricular goal is efficient coverage of the syllabus; his ultimate pedagogical goal is a high passing rate. He even defended this practice when Ruth visited him during office hours. To meet his curricular and pedagogical goals, he needed to underestimate the ability of his students, teaching in a simplified, superficial, linear mode. Students pass his course, but very few learn to be independent or critical thinkers, at least under his tutelage. Ruth, in fact, was his only 'A' student that semester.

Mike earned his tenure in five years, enhanced primarily by his "excellent teaching." He completed his syllabus every semester and mastered the departmental style of lecturing. He has yet to earn promotion, perhaps for various administrative or scholarship reasons (he only had two publications by his sixth year). However, one cannot wonder if his promotion will be ultimately affected by an incomplete dualism relative to his department elders. While Cathy's second semester general chemistry professor was brutally and abusively hegemonic, eventually inhibiting her from seeking office hour help, Mike's door was consistently open for students. While Mike's pedagogy geared toward a high student passing rate, Cathy's professor sought to fail as many students as possible (he once failed 57% of his class), or at least drive as many as possible to drop the course. Dr. A, in fact, uses a similar approach in his organic chemistry course, the course that made him a Distinguished Professor.

Two senior faculty imposing their hegemony with "bad" teaching; Mike, still seeking promotion, only practices "benign" teaching. Perhaps in the eyes of the department elders, he is not "ready" yet for promotion; his capitulation is not complete. Mike cannot be a true master until he perfects his apprenticeship. When Mike completely capitulates, when he finally regresses to monolithic dualism, his ultimate hierarchical reward will come. The elders know exactly why tenure and promotion requires so many years; it takes that long just to "housebreak" new, idealistic professors.

Beverly Gordon (1999) describes academia as "the Hood," a structure based on epistemological domination, where the ultimate paradigm destroys all other paradigms. Because knowledge is cyclical and redundant, it is nearly impossible to break a paradigm's dominance. Because western education and science do not respect alternative views, scholars essentially have three options by which to battle dominance and intolerance: subvert, capitulate, or flee.

Cathy chose to subvert. Ruth, who asked *why bother fighting it?*, may possibly capitulate. Jim ultimately chose to flee. These are university undergraduates, who may have become science educators. They may have become university professors like Mike. Regardless of status—student, teacher, professor, or administrator—we are all confronted with the options to subvert, capitulate, or flee. Those options are always tied to a preordained system of reward and punishment.

I am no longer a faculty member at the university Ruth, Cathy, and Jim continue to attend. Shortly after their first-semester course, I was administratively forbidden from teaching general chemistry for questionable teaching practices. As I continued my research and interacted with students, I was subsequently forbidden from continuing this research, for its lack of departmental relevance. Six months later, I was dismissed from the university for being an incompetent and unproductive member of the chemistry faculty. Consider the following excerpt from my former department's formal recommendation for my dismissal:

> His philosophy is neither properly understood nor accepted by the majority of our department's faculty. We do not think that he is an effective teacher in chemistry courses. We respect his dedication to the students, and his desire to engage them in the learning process, but we feel strongly that he is failing to provide an adequate level of instruction. He is ineffective and ill-qualified as a professor in our department. He is good for the students; they like him and are comfortable learning in his class. But he's not good for our department; the needs of the department come before those of students.

How can we learn about classroom educators' influence—positive or negative—through the eyes of students? Regardless of what educators seek to achieve, students are perceptive enough to sense educators' goals and beliefs. The way we teach and interact with our students can have a profound impact on the way they learn and value science. The postmodern condition inhibits the embrace and celebration of knowledge; we seek a pedagogical and cultural antidote for this.

Every science course we teach has the potential to influence a future teacher. It is true that teachers tend to teach in a manner most familiar to them (e.g., Coppola & Paerson, 1998). What do we want them to be most familiar with? Do we want teachers familiar with low expectations and punitive practices, or do we want them familiar with high expectations and nurturing practices? Do we want teachers familiar with learning goals for long-term growth, or do we want them familiar with archaic objectives that stunt growth? We cannot expect teachers to positively influence students if we don't influence them positively during their academic preparation. The eyes of future teachers are more than tools of observation; they are tools of reflection, mirrors of their learning environment and pedagogical influences.

To conclude this chapter I will share something I overheard while writing this chapter. Our intercultural education committee held a meeting, and the chair informed untenured faculty: *if you're looking for a politically correct committee to serve on, this is it!* The recipes never change. The cookbook remains a classic. The paradigm remains dominant. The hierarchy welcomes capitulators.

> Many students experience artificial success; their attribution and obedience to the arbiters of knowledge reflect this. To become self-reliant is to develop a consciousness of self, an awareness that one can achieve independently and internalize that achievement. Although we strive for advanced cognitive

development, we realize such development cannot take place without an epistemological development. Meaningful epistemologies need to come from within individuals, laying the foundation for future experiences, challenges, successes, and opportunities. We struggle to achieve this goal in the culture of helplessness, but continue nonetheless...we struggle to build a counter-culture of empowerment...building the counter-culture is of great importance. The members within our counter-culture are of great importance. Someday, they will realize this, and transform the culture of helplessness into the culture of empowerment we envision for them. (Pushkin & Colón-González, 1998, p.39)

References

Bartolome, L. (1998, April). The significance of teacher ideology. Presentation at the AERA Annual Meeting in San Diego, CA.

Coffey, A., & Atkinson, P. (1996). *Making sense of qualitative data: Complementary research strategies.* Thousand Oaks, CA: Sage Publications.

Coppola, B.P., & Paerson, W.H. (1998). Heretical thoughts II—On lessons we learned from our graduate advisor that have impacted our undergraduate teaching. *Journal of College Science Teaching, 27,* 416–421.

Driscoll, M.P. (1994). *Psychology of learning for instruction.* Boston: Allyn and Bacon.

Duit, R., & Treagust, D. (1998). Learning in science–From behaviourism towards social constructivism and beyond. In: B.J. Fraser & K.G. Tobin (Eds.), *International Handbook of Science Education* (pp. 3–25). Dordrecht, The Netherlands: Kluwer Academic Publishers.

Freire, P. (1973). *Education for critical consciousness.* New York: Continuum.

Gilbert, J.K., & Boulter, C.J. (1998). Learning Sscience through models and modelling. In: B.J. Fraser & K.G. Tobin (Eds.) *International Handbook of Science Education* (pp.53–66). Dordrecht, The Netherlands: Kluwer Academic Publishers.

Gordon, B. (1999, April). Islands of truth and wisdom: How many epistemologies should there be? Symposium with Cobb, P.A., Collins, A., DiSessa, A., & Packer, M.J. at the AERA Annual Meeting in Montreal, Canada.

Grills, S. (1998). *Doing ethnographic research: Fieldwork settings.* Thousand Oaks, CA: Sage Publications.

Hammer, D. (1995). Epistemological considerations in teaching introductory physics. *Science Education,* 79(3), 393–413.

Hammersley, M. (1992). *What's wrong with ethnography?* London: Routledge.
Have, P. t. (1999). *Doing conversational analysis: A Practical guide.* London: Sage Publications.
Kincheloe, J.L. (1993). *Toward a critical politics of teacher thinking: mapping the postmodern.* Westport, CT: Bergin and Garvey.
Kincheloe, J. L. (1991). *Teachers as researchers: Qualitative inquiry as a path to empowerment.* London: Falmer.
Macedo, D. (1994). *Literacies of power: What Americans are not allowed to know.* Boulder, CO: Westview.
McLaren, P. (1998, April). Hope reinscribed: The struggle for a revolutionary multiculturalism. Presentation at the AERA Annual Meeting in San Diego, CA.
Moll, L.C. (1990). *Vygotsky and education: Instructional implications and applications of sociohistorical psychology.* New York: Cambridge University.
Ollerenshaw, J.A. (1999, March). Opportunities to learn acoustics: fourth grade students meet modern challenges using the oral tradition of storytelling. Paper presented at the NARST Annual Meeting in Boston, MA.
Perry, W.G. (1970). *Forms of intellectual and ethical development in the college years, a scheme.* New York: Holt, Rinehart, and Winston.
Pinar, W.F. (1994). *Autobiography, politics, and sexuality: Essays in curriculum theory, 1972–1992.* New York: Peter Lang.
Pushkin, D.B. (1995). *The influence of a computer-interfaced calorimetry demonstration on general physics students' conceptual views of entropy and their metaphoric explanations of the second law of thermodynamics.* Copyrighted dissertation, Penn State University.
Pushkin, D.B. (1996). A comment on the need to use appropriate scientific terminology in conception studies. *Journal of Research in Science Teaching,* 33(2), 223–224.
Pushkin, D.B. (1997a). Where do ideas from students come from? Applying constructivism and textbook problems to the laboratory experience. *Journal of College Science Teaching,* 26(4), 238–242.
Pushkin, D.B. (1997b). Scientific terminology and context: How broad or narrow are our meanings? *Journal of Research in Science Teaching,* 34(6), 661–668.
Pushkin, D. B. (1998a). Undergraduate science education—Improvement, initiative, and willingness to change. *Journal of College Science Teaching,* 28(1), 8.

Pushkin, D.B. (1998b). Is learning just a matter of tricks? So why are we educating? *Journal of College Science Teaching*, 28(2), 92–93.
Pushkin, D.B. (1999, March). Are students "ready" for constructivism? insights from a laboratory class and beyond. Paper presented at the NARST Annual Meeting in Boston, MA.
Pushkin, D.B., & Colón-González, M.H. (1998, April). Access to knowledge and critical thinking in general chemistry via social constructivism: Pedagogical and curricular opportunities for minority science majors. Paper presented at AERA Annual Meeting in San Diego, CA (ERIC Clearinghouse Paper—Document #ED417-954).
Scheurich, J.J. (1997). *Research: Method in the postmodern*. London: Falmer.
Seale, C. (1998). *Researching society and culture*. London: Sage Publications.
Staver, J. R. (1998). Constructivism: A sound theory for explicating the practice of science and science teaching. *Journal of Research in Science Teaching*, 35(5), 501–520.
Tobias, S., and Tomizuka, C.T. (1992). *Breaking the science barrier: How to explore and understand the sciences*. New York: The College Board.
Valdés, G. (1998). The world outside and inside schools: Language and immigrant children. *Educational Researcher*, 27(6), 4–18.
White, R., & Gunstone, R. (1992). *Problem understanding*. London: Falmer.
Wong, D. (1998, April). Illuminating the humanity and power of science. Paper presented at the AERA Annual Meeting in San Diego, CA.

Chapter Nine
Modernist Traditions in Supervision and the Illusions of Science and Objectivity
Jeffrey Glanz

> I recently became a vegetarian and found beauty and delight in tofu. At nearly the same time, my views of modernist practice of supervision came under scrutiny. For too long I had advocated a modernist approach to supervisory practice that excluded alternate conceptions. In fact, some of my antagonists might gloat if they read this chapter. But this chapter, indeed, marks a turning point in my intellectual understanding of the efficacy of supervision. Even the term itself no longer is as important to me as it once was (J.G.)

Postmodernists have criticized modern conceptions of supervision as bureaucratic, hierarchical, and oppressive. According to a postmodernist view, supervision stifles individual autonomy, especially that of the teacher. A postmodern supervisor seeks to unsettle conventional hierarchical power relationships, replacing such relationships with "relational" ones (Waite, 1997). Anathema is the technicist mindset that imposes preconceived values or notions of "good" teaching through the employment of various supervisory strategies and techniques. For the postmodernist, "the hidden dangers" in "rational-technical thinking are that it reduces supervision to a rigidly defined set of behaviors and responses and places the supervisor in a position to authoritatively diagnose teachers' pedagogical problems and impose particular solutions" (Holland, 1994, pp.11–12). To the postmodern supervisor, the bureaucratic/technicist ontology, fueled by Cartesian dualism, must give way to more holistic, postmodern perspectives.

Advocates of postmodern supervision maintain that supervision should be collegial, nonevaluative, and nondirective. The work of Eisner (1985), Smyth (1991), Garman (1986), Gordon (1992), and Waite (1995) support such a conception of supervision. Moreover, a postmodern view of supervision would seem to even eschew the term "supervision," which in and of itself connotes surveillance and control (Gordon, 1997; Sergiovanni, 1992). Postmodern interpretations clearly favor the term "instructional leadership" as Glickman (1992) explained several years ago:

> Supervision is in such throes of change that not only is the historical understanding of the word becoming obsolete, but I've come to believe that if 'instructional leadership' were substituted each time the word 'supervision' appears in the text, and 'instructional leader' substituted for 'supervisor,' little meaning would be lost and much might be gained. (p.3)

Stephen Gordon (1997) concurs: "My argument is that while the primary goal should be a radical shift from control supervision to collegial supervision, changing the name of what we now call supervision...will increase the chance" that the practice of supervision will change. He too advocates the term "instructional leadership."

I recall writing not too long ago: "I don't think Glickman, Gordon, Sergiovanni, or Starratt (1992) are disingenuous by purposely eschewing the word "supervision," but I do think the penchant for substitute language, in general, is symptomatic of a more widespread trend to speak in euphemisms—sometimes referred to as jargon or educationese. Jerry Pulley (1994), a professor of supervision at the University of Texas-Pan American, in a wonderful little article entitled "Doublespeak and Euphemisms in Education," maintains that our propensity for political correctness or what Lasley (1993) calls "pedagogical correctness," in this context, has beclouded our perspective so much so that our language has become confused and self-contradictory at best and "grossly deceptive" and evasive at worst.

No, the field of supervision has not evolved to a point that it should be called something else. My contention is that supervision should be called what it is—supervision is about engaging teachers, face to face, in an effort to improve instruction with information, techniques, and skills that is likely to have beneficial effects on student learning." (Glanz, 1997a)

How has supervision, as a field of study and practice, evolved to such postmodern advocacy? What aspects of supervision have modernists/postmodernists eschewed? What are the origins of modernist approaches to supervision that rely on science and objectivity? Why did supervisors advocate "scientific" conceptions of supervision? To what extent does supervisory practice today rely on science? Are alternative approaches possible? These are some of the critical questions this chapter will attempt to address.

The Bureaucratization of Supervision: Early Modernist Practices

> When I first wrote this I had no idea of what tofu tasted like (J.G.)

Modernist Traditions in Supervision and the Illusions of Science

Earliest recorded instances of the word "supervision" established the process as entailing "general management, direction, control, and oversight" (Grumet, 1979; Gwynn, 1961). An examination of early records from the colonial period indicates that supervision was synonymous with "inspection." Parenthetically, those scholars who imply that early supervisory practices reflected democratic tendencies, at least as we understand democracy today, misread the evidence. Based on my historical investigations (Glanz, 1998), early supervisory practice was a far cry from democratic.

By the end of the nineteenth century, reformers concerned with undermining inefficiency and corruption transformed schools into streamlined, central administrative bureaucracies with superintendents as supervisors in charge (Elsbree, 1939; Gilland, 1935; Griffiths, 1966; Reller, 1935). Supervision, during this struggle, became an important tool by which the superintendent would legitimize his existence in the school system (Glanz, 1991). Supervision, therefore, was a function performed by superintendents to more efficiently administer schools.

Supervision as inspection became the dominant method of administering schools. Payne (1875), author of the first published textbook on supervision, stated emphatically that teachers must be "held responsible" for work performed in the classroom and that the supervisor, as expert inspector, would "oversee" and ensure "harmony and efficiency." In 1888, James M. Greenwood, a prominent city school superintendent from Kansas, described his practices as a supervisor in evaluating classroom instruction:

> Going into a school, I try to put aside everything like authority, or superiority, and to approach the teacher in a proper spirit of helpfulness. Then, I endeavor to see the school from the teacher's standpoint, and, if necessary, to have the teacher see her school as it appears to me.
>
> What to Do?
> 1. I go in quietly. 2. I watch the teacher and pupils awhile, usually until the novelty wears off. 3. Sometimes I conduct a recitation, with the teacher's permission, and thus bring out points in which she may be deficient; or, simply to test the knowledge that the pupils have of the subject. 4. If suggestions should be made to the teacher, I do so privately, or request her to call after school. 5. Depending on the peculiarities of the teacher, the conversation must be directed in such a way as to benefit her. If the teacher be 'heady,' frequently the most efficacious remedy is to let her alone a few days, and when her room is badly demoralized, help her straighten it out. Of a dozen teachers in a building, no two can be helped precisely alike. I think the question may be put in this form: Given the teacher, the school, the defects; how to improve them?

Signs to look for
1.Common sense. 2.Good health. 3.General scholarship. 4.Critical knowledge of the branches. 5.Order. 6.Ability to manage hard cases. 7.Power to teach. 8. Power to develop thought in the pupils. 9.Routine teaching. 10.'Reciting-post' teaching. 11.Skill in questioning. 12.Skill in fertility of resources. 13.Energy and vigilance properly directed. 14.Pleasant voice. 15.Disposition to antagonize pupils. 16.Power to gain the good-will of children without spoiling them. 17. Disposition to scold and to grumble. 18.Attention to pupils reciting and also to those at their seats. 19.Neatness and cleanliness of room, desks, etc. 20.Ability to secure cheerful and thorough work by the pupils. 21.The tendency to waste time doing nothing laboriously. 22.Variableness of teaching. 23.Steadfastness of purpose in teaching. 24.Disposition to take care of school property. 25. Ventilation of schoolroom, and looking after the children's health. 26.Tact and skill in adapting new methods in teaching. 27.Originality in management and in methods. Sometimes I jot down items that need attention and hand them to the teacher...Very much of my time is devoted to visiting schools and inspecting the work. (pp.519–521)

Greenwood's supervisory methods were typical of the era, relying solely on experience and intuition. Note Greenwood's (1891) posture only three years later: The skilled superintendent, said Greenwood, should simply walk into the classroom and "judge from a compound sensation of the disease at work among the inmates" (p.227). Supervision would soon rely on more perfunctory, administrative, and autocratic methods as well. In 1888 centralization was only just beginning to take hold across the nation. Superintendent-supervisors would soon realize that due to increasing numbers of schools under their jurisdiction they would not be able to supervise as closely as they would have preferred. As a result of industrialization and urbanization, with its effect on increasing the population of school-aged children, as well as a host of other problems facing urban education, supervision became much more inspectional and evaluative. Still, a review of the literature of the period indicates that Greenwood's supervisory methods, which relied on inspection based on intuition rather than technical or scientific knowledge, were widely practiced.

Supervisors using inspectional practices did not favorably view the competency of most teachers. For instance, Balliet (1894), a superintendent from Massachusetts, insisted that there were only two types of teachers: the efficient and the inefficient. The only way to reform the schools, thought Balliet, was to "secure a competent superintendent; second, to let him 'reform' all the teachers who are incompetent and can be 'reformed'; thirdly, to bury the dead" (pp.437–38). Characteristic of the remedies applied to improve teaching was this suggestion: "Weak teachers should place themselves in such a position in the room that every pupil's face may be seen without turning the head" (Fitzpatrick, 1893, p.76). Nineteenth-century supervisors, for the most part, saw teachers as inept. As Bolin and Panaritis (1992) explained: "Teachers (mostly female and disenfranchised) were seen as a bedraggled troop—incompetent and backward in outlook" (p.33).

The practice of supervision by inspection was indeed compatible with the emerging bureaucratic school system. The raison d' être of supervision in the modern period was to achieve quality schooling by eradicating inefficiency and incompetence among the teaching force. These early modern practices were, however, crude. Modern supervision later gained legitimacy in the educational community through the application of the principles of scientific management, advanced first by Frederick Taylor (1911) and later translated into education by Franklin Bobbitt (1913), whose work will be discussed in the next section of this chapter. During this period, various elaborate rating forms were developed to produce efficient, competent teachers (Pajak, 1993b). Note, in the discussion below, that the improvement of instruction was less important than purging the schools of the inept.

Was such thinking—such a belief system—so far off from my views of supervisory practice? After all, the term itself, "supervision," connotes dogmatic, prescriptive, and hierarchical practice.

As educators, we are bound by our perspectives, our unique vantage points. Reality is perceived and understood by our belief systems that are, in turn, based on assumptions gleaned from our experiences or semantic environments. Reality is dependent on our thinking patterns, belief systems, and mindsets, or as Sergiovanni calls them, "mindscapes" (Sergiovanni, 1991, p.41). Our belief systems are intimately connected to the language we use to articulate and communicate meanings (Wittgenstein, 1958), which influence our actions and behaviors. How we think shapes the world in which we live. As Arthur Schopenhauer, German philosopher, once posited, "the world in which a man lives shapes itself chiefly by the way in which he looks at it." How did I look at the world, the school, the classroom? How did I view teachers?

Supervision as a science: The use of rating scales

American education after 1900 was greatly affected by numerous technological advances. During this time the 'efficiency movement' gained considerably throughout American industry. This movement also had important consequences for American education. Throughout urban cities, curtailment of excessive expenditures, and elimination of waste and inefficiency were priorities. As a result of the work of Frederick Winslow Taylor, who published a book in 1911 titled *The Principles of Scientific Management*, 'efficiency' became the watchword of the day. Taylor's book stressed scientific management and efficiency in the workplace. The worker, according to Taylor, was merely a cog in the business machinery, and the main purpose of management was to promote efficiency of the worker. Within a relatively short period of time, Taylorism and efficiency became household words and ultimately had a profound impact on administrative and supervisory practices in schools. The main person who attempted to adopt Taylor's ideas in the schools was Franklin Bobbitt (1913). Bobbitt's work, particularly his discussion of supervision, is significant because his ideas shaped the character and nature of supervision for many years. What he called "scientific and professional supervisory methods" were, in fact, aimed to legitimize control-oriented supervision. Bobbitt's work and the work inspired by his rating schemes further bolstered efforts by supervisors to remove the incompetent teacher. The only difference now was that an air of scientific legitimacy prevailed.

Franklin Bobbitt, then a professor of educational administration at the University of Chicago, applied the ideas espoused by Taylor to the "problems of educational management and supervision." In 1913 he published a work sponsored by the National Society for the Study of Education entitled "Some General Principles of Management Applied to the Problems of City-School Systems." He presented eleven major principles of scientific management as applied to education. Of particular concern, Bobbitt, then an instructor of educational administration at the University of Chicago, firmly held that management, direction, and supervision of schools were necessary in order to achieve "organizational goals." Bobbitt maintained that supervision was an essential function "to coordinate school affairs;..."Supervisory members must co-ordinate the labors of all...find the best methods of work, and enforce the use of these methods on the part of the workers."

Bobbitt was bent on applying scientific knowledge to the classroom. Bobbitt realized the necessity for developing "scales and methods for measuring the educational product so as to determine with at least reasonable accuracy whether the product rises to standard." He was impressed with the work of T. W. Stone and S. A. Courtis and their attempts to develop "scales of measurement" for teachers. "Ordinarily, the teacher, if asked whether his eighth-grade pupils could add at the rate of 65 combinations per minute with an accuracy of 94 per cent, could not answer the question; nor would he know how to go about finding out," explained Bobbitt. "He needs a measuring scale that will serve him in measuring his product" He praised the efforts of Stone and Courtis in developing "practical usable measuring scales for arithmetical ability." These scales are so important, said Bobbitt, because "each teacher can know accurately what is expected of her...." The teacher, continued Bobbitt, can then know if she is a good teacher or not.

The establishment of "definite scales of measurement" also had important uses for administrators and supervisors, said Bobbitt. A superintendent, for example, could glance over the records of a particular school and be able to discern "weaknesses and deficiencies." He can "locate instantly the strong, the mediocre, and the weak teachers." He could also "see at a glance whether building principals are doing a superior grade of work or relatively poor work." In this way, continued Bobbitt, the superintendent would be "able to tell at once where his strong subordinates are and where his weak ones are." Supervisors also may be benefited by these methods, maintained Bobbitt. Special supervisors may, for example, "by glancing over student records would know whether the teacher is securing the full results expected...and whether in doing so she is handling the normal number of pupils." For Bobbitt, the teacher who would be able to bring a greater number of pupils up to the standard in a minimum amount of time would be a good teacher. In other words, these 'scientific methods' would enable supervisors to categorize 'good' and 'bad' teachers and then be "able to provide appropriate methods to remedy the deficiency."

Bobbitt focused on the qualifications of teachers and attempted to raise their "efficiency." Bobbitt considered "teachers weak" and in need of assistance. Bobbitt declared, "One of the large problems of the supervisor is the treatment of weak teachers." Bobbitt insisted that teachers were poorly trained and that teacher training institutions were not producing 'efficient' teachers. Therefore, it is the responsibility of the supervisor to "bring them up to maximum efficiency." However, Bobbitt decried the "subjective character" of supervision based on "frequently mistaken, and always quantitatively indefinite" judgments. Supervisors, he continued, must be equipped with scientific and objective means to measure the efficiency of teachers. "The way to eliminate the personal element from administration and supervision is to introduce impersonal method [sic] of scientific administration and supervision." The rating of teachers, thought Bobbitt, according to the latest principles of science, would raise supervision to the "lofty status it deserved."

As a result of Bobbitt's work, a number of "rating scales" were advocated and developed by supervisors. Supervisors in the early twentieth century were very much interested in utilizing and devising 'rating schemes' to measure 'teacher efficiency.' One of the early attempts to apply Taylor's model to rate teachers was carried out by Joseph S. Taylor, a superintendent from New York City in 1912. He explained that the measurement of teacher efficiency was essential in New York City schools. He indicated the benefits of a rating scale for teachers: "every teacher who accomplishes the task receives a bonus, not in money, but in the form of a rating which may have money value." Teachers, later on, would viciously attack this idea of basing salary or differentials on results of rating. Taylor also conveyed quite clearly what would result from unfavorable ratings: "...those who are unable to do the work are eliminated...." He explained that supervisors are in a unique position to improve instruction through the use of rating schemes. If, warned Taylor, the teacher is inefficient, the supervisor had every right to say "take my way or find a better one."

The use of rating scales was considered to be of enormous benefit. In 1913, William M. Davidson, superintendent from Washington, DC, stated that "it is obvious that, as the tree is to be judged by its fruits, so the teacher is to be judged by the effects he produces in the pupils of his class." Davidson argued that "the worth of a system of rating is to be determined not merely in terms of its accuracy in stating the facts, but almost as importantly in terms of its tendency to improve the quality of the teachers rated." Davidson contended that systems of rating and criticism "cannot effect the impossible; they cannot recreate the teacher. But much they may do in the direction of stimulating cultural and professional growth."

Charles J. Dalthorp, superintendent from South Dakota, concurred that rating is an invaluable aid in supervision. Dalthorp contended that rating is useful for improving teaching by keeping teachers alert and growing. "It is the duty of the supervisor and executive in the modern progressive school to do everything possible to help the teacher rise to her highest level of efficiency." However, to be effective, rating scales had to scientifically developed and applied. "We possess nothing which the great body of teachers is willing to accept as either just, reliable, or workable," said Henry D. Hervey, a superintendent from New York. If rating is to be successful, Hervey thought it "must be raised out of the realm of mere guesswork and personal opinion and placed upon a genuinely scientific basis, which shall be recognized as such not only by teachers who are to be rated, but also by their friends and by the general public."

Several different versions of rating scales appeared in the early twentieth century. One of the early methods for rating "teacher efficiency" was initiated by Edward Elliott of the University of Wisconsin. "The chief purpose of any teaching efficiency scheme," stated Elliott, "is to serve as the means of promoting development and improvement of the individual teacher." He had hoped to eliminate the rating of teachers based on personal and biased accounts. "The science of education has allowed us to devise objective methods for rating teachers." His scale included categories ranging from physical and moral efficiency to social efficiency. Points, from 0 to 10, would be awarded for each category based on the 'observations' of the supervisor.

Another significant attempt to devise a "teacher efficiency rating scale" and one which was widely disseminated throughout the schools was made by Arthur Clifton Boyce, working as a student in the Department of Education at the University of Illinois and later at the University of Chicago, in 1915. Boyce, in devising his method, first conducted a study of 350 cities over 10,000 in population, asking schools to report their methods of rating teacher efficiency. Boyce discovered that most, if not all schools, relied on what he called the "impression method" of supervision. This method included rather impressionistic and subjective conclusions after a brief visitation in the classroom. One administrator in Newburgh, New York, quoted by Boyce, stated that "we have 70 teachers, and our means of judging them is by visiting their classrooms and observing their work." Boyce concluded that this "impression method" used by many schools in rating teacher efficiency was inadequate for a number of reasons. "The weakness of these schemes," charged Boyce, was that they "result from (1) inadequate analysis...(2) a lack of definition of terms, resulting in vagueness and indefiniteness, and (3) the method of recording judgments, which is frequently wasteful of time or inaccurate or uncontrolled...." Boyce claimed that his "scheme" would "overcome to some extent these difficulties by incorporating a comprehensive list of qualities, careful definition of terms, and the graphical method of recording judgments." His scale was composed of 45 different items, grouped in five main headings, as follows: "personal equipment, social and professional equipment, school management, technique of teaching, and results." His scale, like many others during this period of time, ranked teachers in each of these categories using value judgments such as very poor, poor, medium, good, or excellent. The evidence indicates that these scales were used quite extensively in many schools across the nation. The function of supervision continued, during this era and not unlike the past, to be inspectional, albeit with a scientific base.

The application of science to the study and practice of supervision continued unabated throughout the 1920s and 1930s. One of the foremost proponents of science in education and supervision was Alvin S. Barr. He stated emphatically that the application of scientific principles "is a part of a general movement to place supervision on a professional basis." Barr (1931) explained the importance of science in supervision in his widely disseminated book, *An Introduction to the Scientific Study of Classroom Supervision*. Barr stated in precise terms what the supervisor needed to know.

> Supervisors must have the ability to analyze teaching situations and to locate the probable causes for poor work with a certain degree of expertness; they must have the ability to use an array of data-gathering devices peculiar to the field of supervision itself; they must possess certain constructive skills for the development of new means, methods, and materials of instruction; they must know how teachers learn to teach; they must have the ability to teach teachers how to teach; and they must be able to evaluate....In short, they must possess training in both the science of instructing pupils and the science of instructing teachers. Both are included in the science of supervision.

Barr maintained that teaching could be broken down into its constituent parts and analyzed. The success of a science of supervision, Barr asserted, depends upon "breaking up this complex mass into its innumerable elements and to study each objectively."

Another noteworthy proponent of scientific supervision was Charles H. Judd. Judd (1920), in an address before the National Association of Secondary School Principals, stated that "teachers must be supervised in a fashion which is at once direct and scientific." Judd criticized the manner in which supervisors were chosen without adequate training in the science of education. In the future, he said, "They will be selected because they are equipped by mental capacities and by careful scientific study for administrative and managerial functions." Judd urged further that "both the non-supervisory attitude and the attitude of excessive supervision ought to be replaced by scientific method of determining whether classroom work is efficient or not."

Unfortunately, criticism of rating scales emerged shortly after their implementation in the schools. Writing in 1920, H. O. Rugg, then of the Lincoln School at Columbia University, stated that "the movement to rate teachers...needs a new impetus and a new emphasis." Rugg claimed that these "schemes" were "nearly always opposed by the teachers themselves and frequently the [supervisors] have been skeptical of their value." Rugg identified three shortcomings of rating schemes. First, the rating cards in practice "are not aimed at self-improvement" and have frequently been "an administrative scheme superimposed from above." Second, rating schemes, according to Rugg, were biased and abstract. "Rarely have such schemes been made concrete enough so that two or more rating officers rating the work of the same teacher could visualize precisely the same group of qualities." Third, concluded Rugg, the classification of traits themselves was ambiguous and ill defined.

Rugg devised his own rating scale, which was divided into five headings: skill in teaching, skill in mechanics of managing a class, teamwork qualities, qualities of growth and keeping up to date, and personal and social qualities. Rugg used a classification of low, medium, or high. The second part of his scale, Form B, was a numerical scale based on teacher-to-teacher comparisons. Rugg warned educators that no matter what scale of rating is used, ratings should always be done a number of different times in order to get an accurate picture of the teacher's efficiency. "No single rating on teachers should be used as a measure of that teacher's efficiency. Conditions should be found by which at least two administrative officers can rate each teacher. If not," continued Rugg, "the final rating on a teacher should certainly be the average of several independent ratings of the same officer."

A committee of southern educators surveying various rating schemes published their findings in the *American School Board Journal* (1921), under the title "The Rating of Teachers." They concluded that many rating schemes "now in use around the country" had serious defects and deficiencies. However, they stated that despite the defects, "rating must go on." "Defects are to be found in every score card, in order to secure the best results in rating teachers, some mechanical device must be used." Writing in 1922, Franklin W. Johnson, of Teachers College, Columbia University, criticized rating scales "because of their emphasis on qualities of teaching rather than on the results of teaching." He, as well as other educators, decried the use of rating scales as autocratic devices used "to judge and assume to measure the fitness of their teachers for retention or for earning promotion."

Criticism of rating scales, at times, was virulent. One of the most scathing critiques against supervision and rating came from Sallie Hill, a teacher, in 1918. In referring specifically to rating, Hill charged, "there is no democracy in our schools." "Here let me say that I do not want to give the impression that we are sensitive. No person who has remained a teacher for ten years can be sensitive. She is either dead or has gone into some other business," stated Hill. She particularly disliked supervisors who were inadequately trained and incompetent. "It is humiliating and tends to neither cheerfulness nor hopefulness," said Hill, "to have to submit one's work to the criticisms of those whose lack of training and experience has not fitted them for their positions." She concluded by saying that there are "too many supervisors with big salaries and undue rating powers." Six years earlier, an editorial appearing in the *American Teacher*, the journal of the American Federation of Teachers (AFT), stated that "there is probably nothing, not even meager salaries, that frets and worries teachers more than supervision does."

In 1920, Leroy A. King stated quite emphatically that there was no available evidence as to the value of "teacher rating scales." L.J. Brueckner, a noted textbook author in the field of supervision, explained the unreliability of the rating of teachers by supervisory and administrative officers. "In the first place," said Brueckner, "there are lacking any objective standards by which the work of the teacher may be rated...In the second place, ratings are unreliable because they usually consist in a composite rating based on ratings of specific factors which are unweighted as to their importance." He continued, "they are evaluated almost entirely in terms of the subjective judgment of the rater....This results in merely compounding the error of the rating" (National Conference on Educational Method, 1928). Willard S. Elsbree, associate professor of education at Teachers College, Columbia University, stated that "the refinement of our measuring instruments and the development of a clearer concept of what constitutes good teaching," would mean that "the time may not yet come when we can measure teaching efficiency with an accuracy which approaches our present measurements of intelligence." "In the meantime," continued Elsbree, "...there is little if anything to be gained by employing the crude rating devices which are now available."

Despite the criticisms against rating schemes, they were widely used in schools throughout the first twenty-five years of the twentieth century. During this time, many supervisors and professors of educational administration attempted to promulgate their own scales. The proliferation of rating devices contributed too much confusion, however. No agreement was reached regarding uniform standards for measuring teacher efficiency. It was possible and often commonplace for a teacher to be rated unsatisfactorily with one scale, yet competent with another. With no-agreed upon criteria for excellent teaching, college professors and supervisors began to question the efficacy of rating schemes. Criticism also mounted against rating scales because in many instances they were not used to offer any constructive advice as how to improve teaching. Rating scales were merely seen as ways of categorizing, stigmatizing, and controlling teacher behavior. Teachers argued that rating scales were not consistent with the ideals of democratic schooling and only mirrored bureaucratic and autocratic methods of the nineteenth century. One visionary, for example, stated in 1917, "the principal object of supervision should be the improvement of teaching....Rating," he continued, "cannot achieve this objective."

Science in education and in supervision, in particular, took firm hold during the first four decades of the twentieth century. These ideas found justification within bureaucratic schooling and were supported by the beliefs of prominent educators. Modernist supervision theory and practice relied heavily on the science of education and supervision. Scientific supervision would continue throughout the twentieth century. While crude forms of rating scales might have disappeared, attempts continued to "objectify" and find a more "objectively determined" basis to supervise teachers.

In sum, supervision in the modern era, at least initially, was characterized in two ways: by "inspectional practices, which reflected the "emergence of bureaucracy" in education, and by the "social efficiency" movement. The movement to alter supervisory theory and practice to a more scientific focus would not occur until the 1920s as a direct result of growing opposition to autocratic supervisory methods. "The practice of supervision should rely on objective, verifiable data. To do otherwise, would be unprofessional."

The Origins of the Reliance of Science and Objectivity

Supervision, as a role and function, in schools found legitimacy with the emergence of bureaucracy in education coupled with the "social efficiency" movement. Within this context, supervisory practice focused on the application of rating scales as an attempt to objectify and make supervision more scientific. Yet, supervisors realized that growing opposition to their methods of "scientific" supervision necessitated a change.

Supervisors, in their attempt to gain professional recognition, looked toward science as a way of legitimizing their work in schools. Supervisors believed the use of scientific principles in their works with teachers would enable them to gain acceptance as professionals. Unfortunately, their scientific notions were highly speculative and visionary, and as such did not gain acceptance in the public schools. This was due in large measure to their lack of attention to, and understanding of, the ideas of a man whose views of science in education were profound. The work of John Dewey (1929), particularly his work about the "science of education," did not gain enough notoriety among educators in general, and supervisors, in particular. Dewey believed that the future of civilization depended "upon the widening spread and deepening hold of the scientific habits of mind; and that the problem of problems in our education is therefore to discover how to mature and make effective this scientific habit." Dewey held that

> Science must have something to say about what we do, and not merely about how we may do it most easily and economically....When our schools truly become laboratories of knowledge-making, not mills filled out with information-hoppers, there will no longer be the need to discuss the place of science in education....The problem of [an] educational use of science is to create an intelligence pregnant with belief in the possibility of the direction of human affairs by itself....The method of science engrained through education in habit means emancipation from rule of thumb.

Dewey asserted that science, to have any lasting effect in schools, must be grounded in the "lived experience" of the members of each school.

> Science is experience becoming rational. The effect of science is thus to change men's idea of nature and inherent possibilities of experience. By the same token it changes the idea and the operation of reason. Instead of being something beyond experience, remote, aloof, concerned with a sublime region that has nothing to do with the experienced facts of life, it is found indigenous in experience: the factor by which past experiences are purified and rendered into tools for discovery and advance.

Dewey, in sum, believed that scientific theory was related to practice "as the agency of its expansion and its direction to new possibilities."

Dewey's (1929) most scathing critique of existing scientific practices in the schools, as well as the most lucid exposition of his ideas on scientific inquiry, was set forth in his widely read, if not understood, book, *The Sources of a Science of Education*. In response to the questions: "Is there a science of education? Can there be a science of education?", Dewey replied that while scientific and systematic investigation sheds light on a range of facts by enabling "us to understand them better and to control them more intelligently; less haphazardly and with less routine," our current utilization of science is inadequate and misdirected. Dewey denounced the current practice of science in education. There is "a strong tendency to identify teaching ability with use of procedures that yield immediately successful results, success being measured by such things as order of the classroom, correct recitations by pupils I assigned lessons, passing of examinations, promotion of pulls to a higher grade, etc." Supervisors, charged Dewey, "want recipes for classroom success." This view of "science is antagonistic to education as an art," declared Dewey. Dewey claimed the use of rating schemes was not an "enhancement of science in education" but a detraction from the true aims of science. "Such attempts, even when made unconsciously and with laudable intent to tender education more scientific, defeat their own purpose and create reactions against the very concept of educational science." Dewey concluded his little book with a recapitulation and final admonition. The only way, said Dewey, to create a science of education is to involve oneself in the "educational act itself." The intense interaction between practitioner and pupil will in and of itself yield "scientific formulations." "Education is by its nature an endless circle or spiral...in its very process it sets more problems to be further studied, which then react into the educative process to change it still further, and thus demand more thought, more science, and so on, in everlasting sequence." Dewey warned that to ignore the value of "experimentation and discovery" will lead to a mistaken conception of the "true meaning of scientific inquiry." Science based on experimentation, said Dewey, is emancipatory and purposeful.

Dewey's ideas of science as applied to educational practice did not receive much attention. Supervisor, in particular, did not adopt Dewey's model of scientific inquiry. Much of his writing, especially about the science of education, was technical and enigmatic in its presentation. As a result, confusion and misinterpretation of Dewey's views prevailed. Given the fact that there was much misunderstanding it was not surprising that supervisors did not adopt Dewey's ideas. In addition, supervisors eschewed his ideas about science because they were more interested in definite, ready-made prescriptions. Dewey's admonition to avoid definitive scientific formulations in favor of gradual experimentation of ideas in the classroom did not find favor among supervisors.

Supervisors desperately wanted instant solutions to the problems they faced in schools. Rating schemes, for example, were appealing to supervisors because they could, it was thought, accurately assess the performance of teachers' work. In doing so, supervisors hoped to gain respect and stature.

Attempts to alter conceptions of supervision that relied on science proved unsuccessful. Modernist practices held sway, supported by the continuation of bureaucracy in education and schooling. Supervision has continued to be defined and seen as vitally important in the promotion of instructional improvement. Modern conceptions of supervision, therefore, promote professional growth of teachers, foster curriculum development, and support instruction. Modernist supervision provides assistance to teachers and direction for supervisors-in-training whether through clinical practice (Goldhammer, 1969), developmental supervision (Glickman, 1985), cognitive coaching (Costa & Garmston, 1994), or group development and action research (Glickman, Gordon, & Ross-Gordon, 1998). Such task-oriented approaches, supported by the application of descriptive research methods and clinical practice, characterize supervision in the modern era.

Postmodernists, viewing schooling as indeterminate, nonlinear, cyclical, and contingent, criticize modernists who claim that supervision at its best accumulates practically verifiable information (e.g., data that indicates Teacher X is allowing girls less time to respond to questions as compared to boys) so that the teacher can view his classroom from another perspective and consider alternate ways of doing things. Postmodernists, or those with postmodern perspectives, suggest that modernist views of supervision are overly technicist in orientation. As an alternative, they might suggest "dialogic supervision"(Waite, 1995), which advances collegial relationships between supervisors and teachers. Waite, for instance, advocates the "null technique," in which the supervisor becomes "witness to a teaching episode in order to enter into a dialogue with that teacher..." (Waite, 1997). Dialogic supervision seeks to enhance the quality of the teacher-supervisor relationship by focusing more on the dialogue than the "data." This way, says Waite, "both the teacher and supervisor have a better chance of coming to the table on an equal footing...." "Egalitarian reciprocity" is what Waite suggests.

> Why did I alter my conception of supervision? I think the change was inevitable, almost evolutionary. The more I thought about knowledge and learning, the more I worked with student teachers in the classroom, and the more workshops I gave to supervisors, the more I realized the importance of a

more inclusive, broader understanding of what was called supervision.
As I was approaching my late 40s, health became more important to me. I decided to change my eating habits. I found a nutritious food substitute called tofu. Eating tofu and marveling at its versatility made me think....Now I am not really implying that my introduction to tofu (or tofu itself) influenced my newfound conception or belief of supervision (or the other way around). I do think, however, both changes are curious, if not fortuitous. (J.G.)

Three Ways of Doing Supervision

Three approaches to supervision have been suggested by May (1989). These models of supervision may very well reflect thinking and practice of supervision over the last fifty years. The three models of supervision articulated by May (1989) include the Applied Science Approach (which represents a modern conception of supervision); the Interpretive-Practical Approach; and the Critical-Emancipatory Approach (both of which may represent a postmodern perspective).

The Applied Science Approach

This approach to supervision relies on the empirical-analytical sciences and emphasizes technical aspects of the supervision process. At its most basic level, this applied science approach assumes that certain school personnel are in a better position to oversee the instructional process than others. In May's words, "This conception suggests that supervisors are experts and teachers are not [necessarily]....This view of teaching and/or supervision carries several labels which embody a theme of control: directive, executive, behaviorist or positivist...."

Using this approach implies that supervisors diagnose problems in the classroom after a series of close observations. Supervisors then prescribe a particular course of action, and teachers are expected to incorporate the suggested changes. Suggestions offered presumptively are drawn from a research base. Suggestions pertaining to technical classroom management skills and specific teaching strategies are common. Examples of this approach are currently reflected in data collection of teaching via Glickman, Gordon, and Ross-Gordon's (1998) "categorical frequency instruments," or Acheson and Gall's (1987) "selective verbatim" technique. Providing these data, it is thought, affords teachers the opportunity to reflect and view their classroom through "another set of eyes."

The applied science approach is technically oriented, hierarchical in its organizational structure, and most often associated with modern views of supervision. This prescriptive model is often called directive or evaluative supervision. This model dominates supervision discourse in both theory and practice. At least according to nonmodernist views, the illusions of science and objectivity have dominated supervisory practice in schools. Traditions, it is argued, are difficult to dispel, especially as bureaucracy in education and schools prevails.

The Interpretive-Practical Approach
The interpretive-practical approach is reflected in "person-centered" supervision. "Uniform answers to educational problems are viewed as impossible to apply because practical problems are seen to be context bound, situationally determined, and complex." The supervisor is not the overseer or prescriber but a guide, facilitator, or confidante. Relying on enhanced communication and shared understandings, this approach encourages interpersonal and collegial aspects in the supervision process. This model is often called consultative or collaborative supervision. Clinical supervision, embodying neo-progressivism, may, but does not always, characterize this approach (Hopkins & Moore, 1995).

Supervisors with a modernist bent might argue that they too are very concerned with the human relations element of supervision. A modernist supervisor, however, doesn't find evaluative supervision incompatible with the importance of nurturing the human enterprise. Supervision, they assert, represents the process of supporting instructional services, thereby meeting the aims, goals, and objectives of the school organization. Holding teachers accountable and ensuring high-quality instruction are primary goals.

The Critical-Emancipatory Approach
May (1989) believes that this approach encourages reflective action on the part of both teachers and supervisors. Going beyond scientific objectivity and mere collaboration in the development of instructional goals, this approach challenges teachers to "examine the moral, ethical, and political dimensions embedded in everyday thinking and practice." Intending to raise teachers' consciousness and critical awareness of the sociopolitical contexts in which they work, emancipatory supervisors challenge teachers to take risks and construct knowledge for themselves (see, e.g., Bowers & Flinders, 1991; Waite, 1995). This approach to supervision emphasizes diverse educational perspectives to formulate visionary and consensually developed supervisory practices. Critical to this aim is the dissolution of hierarchical frameworks that preclude teacher involvement in supervisory decision-making. Postmodernist approaches to supervision assert that teachers and so-called supervisors

Modernist Traditions in Supervision and the Illusions of Science 231

must continually question the relationship between knowledge and power and be willing to dismantle and replace educational practices that impede efforts to truly improve instruction.

Those who advocate this approach find reliance on science and objectivity not only to be problematic but ill conceived. A postmodern approach to supervision should become a contextually based activity, which provides pragmatic rather than prescriptive conversations as well as in-the-moment deliberation and reflection-in-action. Postmodernists in this context realize the indeterminacy, paradox, and chaos endemic to contemporary classrooms. They argue that teachers should be permitted to deconstruct, reconceptualize, and define their own educational experiences without the strictures imposed by modernist practice. Regarding the application of science in supervision, postmodernists realize that educators must work with certain tensions, chaos, and contradictions that are not fully resolvable. Working comfortably within ambiguity and complexity rather than lusting after certainty and reason is a postmodern response to questions of science and objectivity. Modernist scholarship must move beyond mere adherence to positivist inquiry to supervision work founded on democracy and participative leadership.

In this chapter, thus far, I have discussed supervision, as a field and practice, in its efforts to eschew its bureaucratic and hierarchical "modern" legacy. Postmodern approaches denigrate a reliance on science and objectivity. Moreover, such efforts, it is argued, are illusionary given the postmodern condition in which schools are situated. I have explained the origins of this modern technicist mindset and why supervisors adopted such advocacy and practice. I have indicated that modernist traditions still hold sway in many schools today. Is there an alternative? Before I present a suggested model in which to view supervision more holistically, it should be emphasized that the three models described above should not be viewed as evolutionary in the sense that one replaces the other as individuals make advancements in the supervision field. Rather, all three approaches, the technical, practical, and political, have viability and applicability for instructional improvement. As May (1989) argues, "each framework suggests a legitimate human interest."

> You see, even though I love tofu and now realize the efficacy of postmodern supervision, as an alternative to traditional practice, I still maintain moderation in my eating habits as I do in my understanding of the theory and practice of supervision. Eating tofu has taught me that, as I'll soon explain. (J.G.)

In this last section, I expand upon a metaphor for supervision that, I believe, can help us recast and refocus our thinking about educational supervision that embraces both the modern and postmodern views (for both surely have no monopoly on the "truth") (Glanz, 1997b).

Tofu as a Metaphor for Supervision
Supervision should be conceived as that process which utilizes a wide array of strategies, methodologies, and approaches aimed at improving instruction and promoting educational leadership as well as change. Those concerned with supervision may then work on curriculum development, staff development, schoolwide reform strategies, action research projects, and mentoring, while, at the same time, they may utilize directive, collaborative, or empowering methods. Supervision is supervision regardless of the context in which it is practiced (e.g., preservice and/or inservice settings). Supervision as such does not become meaningless or lack purpose. Rather, supervision is pliable enough to meet a wide range of instructional needs. Remaining responsive to diverse demands would be the field's greatest asset. "Supervision as Tofu" in this context becomes an apt metaphor.

"Tofu," translated into English as "bean curd" or "soybean curd," is an important product of the soybean used in China for more than 2,000 years. Rich in proteins, vitamins, and minerals, low in calories and saturated fats, and entirely free of cholesterol, tofu appears to be the ideal food. Tofu is also unique because it has no taste. Tofu's remarkable quality is that it assumes the flavor of any other food with which it is placed. Tofu can be marinated, stir-fried, scrambled, baked, broiled, grilled, steamed, or barbecued. As Paino and Messinger (1991), authors of *The Tofu Book* state: "It can hide in your cannelloni, taco, or stew, and—before your eyes—take on the flavor of those and many other foods" (p.57). Once only found floating in vats in an Oriental grocery or health food store, tofu now is found in colorful packages on the shelves of many supermarkets.

Tofu's unique quality to remain almost incognito and yet to assume the flavor of its host dish without loss of its nutritional value can be a useful analogy for educational supervision. Supervision is tofu in the sense that it no longer must conform to prescribed or expected practices. Supervision is tofu in the sense that it is flexible enough to represent a wide array of instructional and reform strategies. Supervision is tofu in the sense that, although unseen at times, it remains a supporting service for teachers. As such, supervision as tofu retains its integrity yet remains responsive to diverse demands. Supervision as a function survives and flourishes because it is able to offer instructional assistance amidst a rapidly changing and complex school system.

Supervision is also tofu because its knowledge base is broad, inclusive, and liberal. Supervision thus can function in a variety of settings with diverse groups of teachers, each possessing unique and varied needs. With supervision now broadly conceptualized and practiced, it is not limited to particular methodologies. Supervision can achieve conceptual clarity in this context because its practitioners no longer fear the use of "pedagogically incorrect" strategies when appropriate and warranted. "Directive," "differentiated," "transactional," and "transformational" supervision all find suitable justification within this more encompassing view of the field. Like other fields such as counseling (e.g., Williams, 1995) and religion (e.g., Pohly, 1993), supervision so practiced in schools becomes purposeful, relevant, and influential.

Many examples of "Supervision as Tofu" may be offered. Alternative forms of professional development of teachers represent "supervision as tofu" in action. The following case study (Sullivan & Glanz, 1999) highlights one such course of action.

The International School is one of four mini-schools or Institutes that comprise a large New York City middle school. It was previously one of the lowest performing middle schools in the district. Nancy Brogan, an assertive, go-getter principal, was brought in to try and improve both achievement and image. Open to innovation and aggressive in pursuing funds, she broke up the 1,200+ students into four theme institutes. The International Institute, one of the four mini-schools, is composed primarily of Haitian, Russian, Spanish, Chinese, Bengali, and Urdu bilingual students. Because of the bilingual focus, the staff of the Institute mirrors the diversity of the student body.

Through one of Ms. Brogan's outreach efforts, she secured the assistance of Susan Sullivan (my co-author, see previous reference). Susan's initial project was to help organize the governance committee of the Institute. This task completed, conversations veered more towards curriculum and teaching issues. All of the teachers on the steering committee were committed, enthusiastic, effective and creative teachers, and along with the Institute director Lynn Pagano, were open to anything that would promote student achievement. Since peer coaching was an approved choice in the new union contract's weekly period for professional development for each teacher, it was decided to pursue the possibility of using this staff period to develop and implement the skills and practices of peer coaching. A second recently approved union provision included the option of alternative assessment for tenured teachers. The director's reaction was: "These teachers are the best. They can all use the peer coaching as their official observations—even the one who is just shy of tenure. This way I can concentrate on the teachers who really need help. By the way, would you mind if I sat in on your meetings?"

The prospective participants and Susan then made a site visit to a school that had developed a very sophisticated system of peer assessment. The teachers came back excited and ready to take on the challenge.

Next they had to decide what was going to be the focus of the peer coaching. Two of the teachers had been trained during the summer in the new standards that the city and state had begun to require and one had been involved in developing the Spanish curriculum and city adaptation for the standards. Their enthusiasm about the work they had been doing and the need for implementation triggered the conversation about two possible coaching models: 1) Peer observations based on the implementation of the curriculum for the new standards. 2) Coaching where teachers would discuss classroom challenges or interests and conduct interclass visitations.

Mannor Wong, the Chinese bilingual teacher commented: "Since I'm not tenured yet, I'd prefer honing my general instructional techniques. I'd like someone to be able to give me feedback on the strategies I'm working on in the classroom and how the students are responding to them."

Farouki Naserin, the Urdu bilingual teacher, had the following conversation with Madeline, the Spanish bilingual teacher: "Since you've already developed curriculum in Spanish for the new standards, could I see how you're going to implement it in the classroom? Then maybe you could observe me as I try to use the adapted curriculum in my Urdu classes?"

"That's a good idea. And the students will feel comfortable with you there because of the multicultural sharing we've been doing."

Lynn Lavner, an ESL teacher with a strong performing arts background, began unofficially sitting in on our meetings. Her reactions were somewhat different: "I could use help as I try to develop our South Pacific curriculum this spring, but I don't know about observing another teacher's class. I don't think that's for me."

The plan that emerged was for the participants to learn and practice interpersonal, observation, and feedback skills through observations of videotaped classroom instruction and role-plays of the interpersonal and feedback approaches. They would then be prepared to help each other more effectively and become turnkey trainers for future coaching groups. A date for the first orientation and training meeting was set. In the interim, the volunteer group was to be finalized.

The Best Laid Plans of Mice and Teachers

They began meeting in the Director's office during the teachers' 45 minute lunch hour. There were constant interruptions, time lost getting lunch, and teachers arriving late or not at all. Since Lynn, the performing arts ESL teacher, hadn't been in on the original planning, her initial attendance was spotty. Mannor had not been involved in the early meetings either and may have had some initial apprehensions. Susan decided that they needed to go back to the drawing board to look for a longer block of time at a different point in the day. Luckily, this particular group was involved in implementing a grant with some flexible funding. The coordinator of the grant, Elke Savoy, one of the participants, figured out that she could compensate the teachers to come after school for a series of two-hour workshops.

They had one more setback before launching the after-school workshops. Through her ongoing outreach efforts, Ms. Brogan had procured additional professional development help to increase achievement scores. One strategy included daily brief observations by the directors in all classrooms with completion of checklists for each teacher. All staff was to follow certain procedures that the directors would verify in their visits. Susan met with the principal and Ms. Pagano to explain that this method was at odds with the peer coaching goals. They agreed that the teachers involved in our project would be exempt from this requirement. In fact, Ms. Pagano was relieved because she was "going crazy" trying to keep up with the marathon-visiting schedule.

Uninterrupted quality time, snacks, and compensation were a few of the elements that fostered time on task. The following weeks were spent perfecting their interpersonal and feedback skills, and using various techniques to observe videos of teachers and students. As they simulated and role-played the skills in class, they began to practice observing colleagues' classes. They finally went through the clinical observation cycle with each other and other volunteers in the International Institute. Discussions and feedback took place at each session to fine tune the process as we went along. Open communication about reactions to the process was encouraged.

Once the participants were comfortable with their observing and feedback skills, the next step was to set up individual or paired plans for their dialogues around curriculum implementation. The initial plans were set up at the last workshop because it was the last opportunity for quality time. They decided that each teacher or pair would write up an action plan for his or her focus. Brief meetings would take place every two weeks to share experiences, provide feedback on what was and was not working, troubleshoot, and modify plans as needed.

Susan kept a running record of the process and provided both a qualitative and quantitative evaluation at the end of the semester. The participants were so enthusiastic that it was decided to involve more volunteers the following fall and share their experience with another institute.

Analysis

Peer coaching is an umbrella term for the many different configurations of teachers helping teachers that have emerged primarily since the 1980s. Some of the other terms often used interchangeably with peer coaching are peer assistance, collegial coaching, technical coaching, cognitive coaching, challenge coaching, peer supervision, etc. Most of these models pertain to variations of peer-to-peer assistance of equals and do not involve evaluation. In this case, peer coaching is defined as teachers helping teachers reflect on and improve teaching practices and/or implement particular teaching skills needed to implement knowledge gained through staff or curriculum development. Some of the major outcomes of peer coaching are

- refinement of teaching practices,
- stimulation of self-initiating, autonomous teacher thought,
- improvement of school culture,
- increased collegiality and professional dialogue, and
- sharing in the implementation of new or common instructional skills

Through the ongoing discussion of teaching and learning, curriculum development and implementation, peer coaching became the heart of professional development at the school. It encompasses all of the skills deemed essential for postmodern supervisory leadership in the 21st century: collaborative relationships, participatory decision making, reflective listening and practice, and teacher self-direction—with the clearly expressed goal of developing autonomous professionals. Without a "Supervision as Tofu" model, if you will, such alternative conceptions would be unlikely to find their way into practice.

Supervision as tofu is diverse and versatile, yet uniform and substantial (like yin and yang). If diversity represents adaptability and flexibility in a range of settings and needs, then tofu is an apt metaphor to describe the work of supervision in schools. Supervisory practice in a postmodern, indeterminate world should proffer alternative models or ways of teaching or doing supervision. Reliance solely on scientific conceptions of supervision is not only misplaced but subversive to the ideals of inquiry and true collaboration. Conceiving supervision as collaborative rather than hierarchical, dialogic versus didactic, descriptive rather than judgmental, and supportive rather than punitive should dominate discourse in the field. As tofu, unassuming yet nutritious, makes an ideal substitute for high-calorie foods, supervision as tofu also blends into the educational landscape to help provide needed services and assistance to teachers.

References

Acheson, K. A., & Gall, M. D. (1987). *Techniques in the clinical supervision of teachers*. New York: Longman.

American School Board Journal.(1921).The rating of teachers, 63:41–45.

Balliet, T. M. (1894). Discussion of Anderson's paper. *NEA Proceedings*, 437–438.

Barr, A. S. (1931). *An introduction to the scientific study of classroom supervision*. New York: D. Appleton and Company.

Bobbitt, F. (1913). Some general principles of management applied to the problems of city-school systems. *Twelfth Yearbook of the National Society for the Study of Education, Part I*.

Bolin, F., & Panaritis, P. (1992). Searching for a common purpose: A perspective on the history of supervision. In C. D. Glickman (Ed.), *Supervision in transition* (pp. 30–43). Alexandria, VA: ASCD.

Bowers, C. A., & Flinders, D. J. (1991). *Culturally responsive teaching and supervision*. New York: Teachers College.

Boyce, A. C. (1915). Methods for measuring teachers' efficiency. *Fourteenth yearbook of the National Society for the Study of Education, Part II* (pp. 9–81). Chicago: The University of Chicago.

Costa, A., & Garmston, R. (1994). *Cognitive coaching: Approaching renaissance schools.* Norwood, MA: Christopher-Gordon Publishing.

Dalthorp, C. J. (1932). Shall we rate teachers? *Educational Method, XII,* 78-81.Davidson, W. M. (1913). How to measure the efficiency of teachers. *National Educational Association Proceedings,* 51, 286-287.

Dewey, J. (1929). *The sources of a science of education.* New York: Liveright.

Eisner, E. W. (1985). *The educational imagination: On the design and evaluation of educational programs.* New York: Macmillan.

Elliott, E. C. (1910). *Tentative scheme for the measurement of teaching efficiency.* Madison, n.p.

Elsbree, W. S. (1939). *American teacher: Evolution of a profession in a democracy.* New York: American Book Company.

Fitzpatrick, F. A. (1893). How to improve the work of inefficient teachers. *NEA Proceedings.*

Garman, N. B. (1986). Reflection, the heart of clinical supervision: A modern rationale for practice. *Journal of Curriculum and Supervision,* 2(1), 1–24.

Gilland, T. M. (1935). *The origin and development of the power and duties of the city school superintendent.* Chicago: The University of Chicago Press.

Glanz, J. (1991). *Bureaucracy and professionalism: The evolution of public school supervision.* New Jersey: Fairleigh Dickinson University.

Glanz, J. (1997a). Has the field of supervision evolved to a point that it should be called something else? No. In J. Glanz & R. F. Neville (Eds.), *Educational supervision: Perspectives, issues, and controversies* (pp. 124–133). Norwood, MA: Christopher-Gordon Publishers.

Glanz, J. (1997b). The tao of supervision: Taoist insights into the theory and practice of educational supervision. *Journal of Curriculum and Supervision,* 12, 193–211.

Glanz, J. (1998). Histories, antecedents, and legacies: Constructing a history of school supervision. In G. R. Firth & E. Pajak (Eds.), *Handbook of Research on School Supervision.* New York: Macmillan.

Glickman, C. D. (1985). *Supervision of instruction: A developmental approach.* Boston: Allyn and Bacon.

Glickman, C. D. (1992). Introduction: Postmodernism and supervision. In C. D. Glickman (Ed.), *Supervision in transition* (pp. 30–43).

Alexandria, VA: ASCD.

Glickman, C. D., Gordon, S. P., & Ross-Gordon, J. M. (1998). *Supervision of instruction: A developmental approach.* Boston: Allyn and Bacon.

Goldhammer, R. (1969). *Clinical supervision: Special methods for the supervision of teachers.* New York: Holt, Rinehart, and Winston.

Gordon, S. P. (1992). Paradigms, transitions, and the new supervision. *Journal of Curriculum and Supervision,* 8, 62–76.

Gordon, S. P. (1997). Has the field of supervision evolved to a point that it should be called something else? Yes. In J. Glanz & R. F. Neville (Eds.), *Educational supervision: Perspectives, issues, and controversies* (pp. 114–123). Norwood, MA: Christopher-Gordon Publishers.

Greenwood, J. M. (1888). Efficient school supervision. *NEA Proceedings,* 519–521.

Greenwood, J. M. (1891). Discussion of Gove's paper. *NEA Proceedings,* 19, 227.

Griffiths, D. E. (1966). *The school superintendent.* New York: The Center for Applied Research in Education.

Grumet, M. (1979). Supervision and situation: A methodology of self-report for teacher education. *Journal of Curriculum Theorizing,* 1, 191–257.

Gwynn, J. M. (1961). *Theory and practice of supervision.* New York: Dodd, Mead and Company.

Hervey, H. D. (1921). The rating of teachers. *National Educational Association Proceedings,* 59, 823-826.

Hill, S. (1918). Defects of supervision and constructive suggestions thereon. *NEA Proceedings,* 347-350.

Holland, P. E. (1994). *What do we talk of when we talk of supervision?* Paper presented at the annual meeting of the Council of Professors of Instructional Supervision (COPIS).

Hopkins, W. S., & Moore, K. D. (1995). Clinical supervision: A neo-progressive approach. *The Teacher Educator,* 30, 31-43.

Johnson, F. W,. (1922). The supervision of instruction. *The School Review,* xxx, 123–129.

Judd, C. H. (1920). The high school manager. *National Association of Secondary School Principals,* 30–31.

King, L. A. (1920). The present status of teacher rating. *American School Board Journal,* lxx, 44–46.

Lasley, T. J. (1993). Rx for Pedagogical correctness: Professional correctness. *The Clearing House,* (67) 77-79.

May, W. T. (1989). Supervision and curriculum. In *Encyclopedia of Education*, 728–730.

National Conference on Educational Method (1928). *Educational supervision: A report of current views.* New York: Teachers College, Columbia University.

Paino, J., & Messinger. L. (1991). The tofu book: The new American cuisine. New York: Avery Publishing.

Pajak, E. (1993). *Approaches to clinical supervision: Alternatives for improving instruction.* Norwood, MA: Christopher-Gordon.

Payne, W. H. (1875). *Chapters on school supervision: A practical treatise on superintendency: Grading; arranging courses of study; the preparation and use of blanks, records and reports; examination for promotion, etc.* New York: Van Antwerp Bragg and Company.

Pohly, K. (1993). *Transforming the rough places: The ministry of supervision.* Dayton: Whaleprints.

Pulley, J. L. (1994). Doublespeak and euphemisms in education. *The Clearing House.* (67) 271-273.

Reller, T. L. (1935). The development of the city superintendency of schools in the United States. Philadelphia: Author.

Rugg, H. O. (1920). Self-improvement of teachers through self-rating: A new scale for rating teachers. *The Elementary School Journal,* xx, 674–686.

Sergiovanni, T. J. (1991). *The principalship: A reflective practice perspective.* Boston: Allyn and Bacon.

Sergiovanni, T. J. (1992). Moral authority and the regeneration of supervision. In C. D. Glickman (Ed.), *Supervision in transition* (pp. 30–43). Alexandria, VA: ASCD.

Smyth, J. (1991). *Teachers as collaborative learners: Challenging dominant forms of supervision.* Philadelphia: Open University Press.

Sullivan S., & Glanz, J. (1999). *Supervision that improves teaching: Strategies and techniques.* Thousand Oaks, CA: Corwin.

Starratt, R. J. (1992). After Supervision. *Journal of Curriculum and Supervision,* (8) 77-86.

Taylor, F. W. (1911). *The principles of scientific management.* New York.

Taylor, J. S. (1912). *Measurement of educational efficiency. Educational Review,* xliv, 350–351, 359.

Waite, D. (1995). *Rethinking instructional supervision: Notes on its language and culture.* London: Falmer.

Waite, D. (1997, March). Super(postmodern)vision. Paper presented at the Annual Conference of the American Educational Research Association, Chicago.

Williams, A. (1995). *Visual and active supervision: Roles, focus, technique.* New York: W.W. Norton & Company.

Wittgenstein, L. (1958). *Philosophical investigations.* Oxford: Basil Blackwell.

Part Three
Pedagogies of the Cultural Studies of Science
John A. Weaver and Karen Anijar

> We can easily see on which side obscurantism and nihilism are lurking when on occasion great professors or representatives of prestigious institutions lose all sense of proportion and control; on such occasions they forget the principles that they claim to defend in their work and suddenly begin to heap insults, to say whatever comes into their heads on the subject of texts that they obviously have never opened or that they have encountered through a mediocre journalism that in other circumstances they would pretend to scorn.
> (Jacques Derrida, 1983, p.15)

The commentaries on the science debates from noted scientists continue to flood publishers' desks. In these commentaries, there is a tendency for scientists to declare that relativists are denying the existence of an objective reality or that anything other than sociological or political issues play a role in the construction of science. After making such declarative statements, these commentators provide no evidence or examples of who exactly is proclaiming the nonexistence of reality. If there is an attempt to present substantive evidence of a relativist conspiracy against the "fools" who actually believe a physical reality exists, this evidence is often taken from comments made three decades ago or more. It is hard to believe that if anyone associated with the cultural studies of science were to make such unsubstantiated or decontextualized comments, they would not be condemned for their lack of scholarly rigor. Yet, scientists and publishing houses tolerate such low standards of scholarship when it comes to critiquing the cultural studies of science. Let me provide two examples from Nobel Prize winners.

In a 1996 review of the Sokal "prank" *The New York Review in Books* Steven Weinberg (1996a) proclaims the worthiness of Sokal's acts and the poverty of French philosophical thought. As an example of his statement, Weinberg (1996a) suggests that Jacques "Derrida and other postmoderns do not seem to be saying anything that requires a special technical language, and they do not seem to be trying very hard to be clear" (p.11). The evidence he supplies to dismiss Derrida's importance to late twentieth century thinking is a statement Derrida made at a conference in 1966 as a response to a question raised by Jean Hyppolite, all of which was later transcribed. To make matters worse, this quote Weinberg uses to dismiss Derrida and other "postmoderns" comes from Sokal's prankster article in *Social Text*. After being called to task for his overgeneralization of Derrida's work, Weinberg (1996b) responded to his critics by concluding: "It seems to me that Derrida in context is even worse than Derrida out of context" (p.56). What we see here is a spiraling downward of academic standards in the name of science; a *National Enquirer*-type approach to constructing and contextualizing evidence in which third-hand accounts are acceptable grounds to dismiss the seriousness of a person's ideas.

Weinberg's comments are problematic for numerous reasons. First, Weinberg labels Derrida a postmodernist, when, in fact, in all of his writings Derrida never describes himself as a postmodernist. If we define, in part, a postmodernist as someone who, within linguistics, argues that the sign has been disconnected from the signifier and signified and free floating signifiers are the realm of meaning (a definition often applied to postmodern thought), then Derrida is certainly not a postmodernist thinker (see Derrida, 1997, pp.38–39). In fact, by these standards of defining postmodern thought, Derrida is a conservative thinker (not in a political sense but in an evidential and intentional sense of reading and interpreting a text), who extends modern linguistics and philosophy rather than abandon, them. If anything, Derrida is a philosopher whose radical thoughts present themself in the form of close readings of texts that are worthy of our intellectual attention. This is a point that Weinberg and Sokal, in their perusals of Derrida's ideas, obviously missed.

Second, by proclaiming that Derrida does not present any special theory that warrants a "special technical language," Weinberg with one broad stroke erases the French and American structural and poststructural traditions and reduces them to incoherent dribble that is neither complex nor insightful. Weinberg assumes that anyone can pick up a book by Saussure, Jacobson, Peirce, or Derrida and understand the contribution they made to our understanding of Western philosophy or Western linguistic theories without any special effort or dedication to understanding the language and culture embedded in these theories. No wonder Weinberg and Sokal view Derrida as an unimportant thinker in Western philosophy; they simply refuse to recognize the importance of understanding that any intellectual theory not only relies on language to communicate its ideas but also lives within a language that loses and gains meaning as time progresses.

Third, the most problematic dimension of Weinberg's comments is his refusal to apply the same rigorous standards that he suggests the cultural studies of science lack. By citing one transcribed comment from Derrida or citing no specific scholar who claims the end of the objective world, Weinberg displays the scholarly standards that we have come to maybe expect from an undergraduate student or a high school student who grafts a quote from the Internet. The issue here is not so much that Weinberg tries to get away with such shoddy scholarship, but why so many of his colleagues within science openly embrace such an approach. Does a Nobel Prize in science confer onto the winners the right to ignore the standards that earned them the respect of their peers?

The second example I want to draw upon to discuss the continual criticism of the cultural studies of science is found in Freeman Dyson's (1999) recent work. Dyson's work is different from Weinberg's in that he is willing to dig a little deeper into the cultural studies of science. Dyson at least admits that there are thoughtful ways of challenging Sokal's assertions and his characterization of the cultural studies of science. In a footnote Dyson (1999) graciously acknowledges that "anyone who reads Sokal's parody should also read Andrew Ross, 'Reflections on the Sokal Affair'...a thoughtful response by one of the editors of *Social Text*" (p.120). Dyson recognizes that reading the works of those being parodied is a worthy academic endeavor. Nevertheless, Dyson demonstrates in the main text of his work, *The Sun, the Genome, and the Internet*, the same proclivity to overgeneralize and cite no examples to provide substance to his claims.

For instance, Dyson breaks up those interested in the history and philosophy of science into two groups: those who follow Kuhn's focus on ideas and those who follow Peter Galison's focus on instruments to understand the movement and development of science. Kuhnians, which Dyson (1999) rightfully points out is a term that does not, and need not, include Kuhn, believe "that the acceptance of new scientific theories is based on political and economic struggle rather than on scientific evidence," and "science can be understood as a social construction without paying much attention to its subject matter" (pp.15–16). Such an approach, Dyson (1999) insists, not only applies to physics but also to biology where "some less perceptive social analysts [comparison here is to the more perceptive Emily Martin, who has been marked by Gross and Levitt as a radical feminist and antiscience advocate, and her work on the popular constructions of the human immune system] have attempted to show that a virus is a social construction" (p.19).

Within the context of each of these comments, Dyson refuses to provide a concrete example of who these social constructivists or relativists might be and what exactly they wrote to give Dyson the impression that they were somehow denying that objective reality existed;

that science was not about objective realities but politics; or that social constructivists were claiming in an either/or manner that science cannot be simultaneously political and scholarly; therefore, politics must be supreme. Dyson provides no evidence to support his construction of the cultural studies of science simply because, like Weinberg and many others in science, he is the real victim of the Sokal prank. Dyson and Weinberg, with all their Nobel hardware sitting somewhere for others to admire, bought into Sokal's reading of the cultural studies of science and concluded that Sokal's reading was not a reading but a reflection of truth; therefore, they did not have to bother reading the very people Sokal parodies. Nor did they feel inclined to ask critical, but fundamentally simple, questions such as "do the cultural studies of science scholars see themselves the same way Sokal or Gross and Levitt interpret them?" It is not Ross and Robbins who have been duped. In reality (if there were a reality), it has been the scientists who have fallen prey to the prank-like scholarship of Sokal, the apocalyptic readings of Gross and Levitt, and the anti-interdisciplinary proclamations of Holton.

In this part, the authors of these chapters are not relativists denying the existence of an objective reality nor constructing a hieroglyphic style of prose. We are interested in demonstrating that if we are to construct a sustainable and "realistic" image of science, the language scientists utilize, the dilemmas of reductionist and determinist thinking, the metaphors they construct to explain their "findings," and the denials they invent in order to protect their modest witness ideology, have to be challenged. This challenge ultimately requires that the cultural studies of science must construct pedagogical strategies so the political, literary, economic, social, and cultural dimensions of science are stressed in the science classroom as much as the intellectual achievements of science. Each of the chapters below offers its own pedagogical vision for the cultural studies of science.

Noel Gough's attempts to invent a pedagogy of the cultural studies of science through the construction of intersecting "(crash) zones." It is in the "(Crash) zone" where science, science fiction, politics, literary tropes and metaphors morph to form a dynamic way of teaching science as all the worldly consequences of doing science are presented to high school students. Gough's "(crash) zone," among other things, challenges a traditional approach to science that is too often reductionist and deterministic. As Gough demonstrates, such a traditional approach is dangerous in our current world of DNA manipulation and cloning experiments. These new scientific developments cannot be explained or constructed within a reductionist and deterministic approach to science. A pedagogy of the cultural studies of science needs to construct an alternative in which students come to realize and see the dangers of thinking simplistically about our DNA, cloning, and other nonreductionist scientific developments. Gough's chapter offers an alternative pedagogy that does not give in to the temptation of traditional science.

In her chapter, Annette Gough offers another alternative pedagogy to traditional approaches. For Gough, the goal for creating pedagogical alternatives in the science classroom is not based primarily in political agendas but rather is founded on a need to construct an approach to science that insures that the world can reap the benefits of science and survive. Gough's route to constructing a pedagogy of the cultural studies of science travels through feminist theories and poststructuralism. In particular, her journey wraps around Sandra Harding's ideas of strong objectivity and postcolonial science. Although much maligned, Gough demonstrates that once we begin to read Harding's work beyond a few quotes here and there and in more depth, Harding encourages us to rethink our notions of objectivity and situated knowledges that influence the way science is done and taught. At stake is not just a careful reading of Harding, as well as other feminist and poststructural writers, but a science pedagogy that respects all peoples and forms of knowledge and contributes to the creation of a democratic world.

It is our goal in constructing pedagogies of the cultural studies of science that as a world and as educators we will begin to break from serious flaws that modern science encourages and promotes through its reductionist and deterministic logic. Contrary to the critics of the cultural studies of science, we are not interested in abandoning science. We are interested in banishing what Bruno Latour (1999) calls a brain-in-a-vat approach to science. The brain-in-a-vat approach simultaneously isolated science from the world while it constructed models for the world. This approach exempts science from being responsible for that which it creates. It encourages citizens of the world from seeing science as a "natural" portal to the world and discourages them from seeing science as a human portal that represents all the creativity and destructiveness that humans can invent. In place of this brain-in-a-vat approach, pedagogies of the cultural studies of science are attempts to move toward the creation of students who are constructed as "bumptious technoscience actor[s]" (Haraway, 1997, p. 94), where science is not dismissed nor worshipped but utilized to construct a democratic and sustainable world. We hope you enjoy these essays and build upon them.

References

Derrida, J. (1983). The principle of reason: The university in the eyes of its

pupils. *Diacritics* (Fall), pp.3–20.
Derrida, J. (1997). *Of grammatology*. Baltimore: Johns Hopkins.
Dyson, F. (1999). *The sun, the genome, and the Internet: Tools of scientific revolutions*. New York: Oxford.
Haraway, D.(1997) *Modest_witnesss@second_millennium.female man©_meets_oncomouse ™* New York: Routledge.
Latour, B. (1999). *Pandora's hope: Essays on the reality of science studies*. Cambridge: Harvard.
Weinberg, S. (1996a). Sokal's hoax. *The New York Review of Books*. (43:13), pp.11–12, 14–15.
Weinberg, S. (1996b). Steven Weinberg replies. *The New York Review of Books*. (43:15), pp.55–56.

Chapter Ten
Teaching in the *(Crash)* Zone: Manifesting Cultural Studies in Science Education
Noel Gough

Why *Crash*?

In this chapter I argue that the mode of storytelling usually known as "science fiction" has special significance for developing appropriate pedagogies of/for cultural studies of science. As Donna Haraway (1989) explains, the term 'science fiction' has in recent years been displaced by a more embracing (and more ambiguous) term, "SF":

> In the late 1960s science fiction anthologist and critic Judith Merrill idiosyncratically began using the signifier SF to designate a complex emerging narrative field in which the boundaries between science fiction (conventionally, sf) and fantasy became highly permeable in confusing ways, commercially and linguistically. Her designation, SF, came to be widely adopted as critics, readers, writers, fans, and publishers struggled to comprehend an increasingly heterodox array of writing, reading, and marketing practices indicated by a proliferation of 'sf' phrases: speculative fiction, science fiction, science fantasy, speculative futures, speculative fabulation. (p.5)

My title deliberately includes a parenthetic reference to an exemplary and germinal SF text, J.G. Ballard's (1985 [1973]) *Crash*, which Scott Bukatman (1993) describes as a "high-tech porn novel" (p.6) and a "compendium of atrocities" (p.326). Bukatman (1993) also describes *Crash* as

> a brilliantly ironic work, set in a postindustrial landscape of highways and automobiles, high rises and airports, television sets and billboards. It is a landscape in which the erotic is denied, repressed and paved over by layers of concrete, tarmac and chrome. Vaughan, the 'hero' of *Crash*, fantasizes his death in a headlong collision with Elizabeth Taylor. The narrator, 'Ballard,' is ever-increasingly drawn toward Vaughan's vision of transcendent sex and violence. (p.292)

I suspect that many readers may not perceive an immediate or obvious connection between science education and a novel in which most characters are obsessed with the sexual connotations of wounds caused by car crashes, so I should state from the outset that the significance of *Crash* for this chapter lies chiefly with its history and narrative form rather than with its thematic content. Istvan Csicsery-Ronay (1991a) nominates Ballard's Introduction to the French edition of *Crash* as "the *de facto* founding manifesto of postmodernist SF" (p.306), and the novel itself has influenced several literary and cultural theorists, including Jean Baudrillard (1991).

I work from a position that takes the generativity of metaphor for granted. Metaphors matter, and they matter in the literal sense that they have material effects. As Norman Fairclough (1992) writes, "when we signify things through one metaphor rather than another, we are constructing our reality one way rather than another. Metaphors structure the way we think and the way we act, and our systems of knowledge and belief, in a pervasive and fundamental way" (p.195). In this chapter, I present "the *(Crash)* zone" as a generative metaphor that structures some of the ways I think and act as a science educator. I recognize that this metaphor may be idiosyncratic, but I trust that by the end of this chapter it will not appear to be too esoteric. I also hope that it might be adaptable. *Crash* is not the only fiction that could be used to modify the concept of 'the zone' that I outline in the following section. For example, *The X-Files* is surreal enough for my purposes, but its overtly paranoid politics means that it lacks the cool detachment of *Crash's* hyper-realized world.

Curriculum Theorizing in 'The *(Crash)* Zone'

If I followed Michel Foucault (1970 [1966]), I might have titled this chapter "Teaching in heterotopia," because I imagine myself teaching science in an "impossible" discursive space characterized by "the disorder in which fragments of a large number of possible orders glitter separately... without law or geometry" (p.xviii). But I prefer to represent the space in which I perform curriculum inquiry as the 'zone' that Brian McHale (1992, pp.250–51), Scott Bukatman (1993, pp.163–82), and a number of other literary scholars identify as a pervasive trope in postmodernist fiction, a site in which multiple worlds are projected and ontological shifts are enabled by fragmenting, reassembling, and/or imploding familiar spaces. Many readers of my generation will recall the portentious words that opened each episode of a well-known 1960s television series:

> You are traveling to another dimension, a dimension not only of sight and sound but of mind—a journey into a wondrous land whose boundaries are that of imagination. Your next stop: "The Twilight Zone".

Teaching in the (*Crash*) Zone

There are many other examples. In the space-age mythology of William Burroughs's (1964; 1966 [1961]; 1967 [1962]) apocalyptic Nova trilogy, the "Interzone" is the region in which anything is permitted and everything coexists. The reconfigured Germany of Thomas Pynchon's (1973) novel, *Gravity's Rainbow*, is called "the Zone," as is a part of Canada mysteriously transformed by alien visitation in Arkady and Boris Strugatsky's (1978) *Roadside Picnic* (filmed by Andrei Tarkovsky as *Stalker* in 1979). "Zones" are also the names that Doris Lessing (1980) gives to the parallel dimensions of the cosmos she creates in her "Canopus in Argus: Archives" series, explicitly introduced in the second novel of the sequence, *The Marriages Between Zones Three, Four and Five*.[1]

While the name may be absent, spaces which serve similar functions to these zones often appear in postmodernist fiction. These include the burgeoning evocations of cyberspace, such as the virtual "Metaverse" of Neal Stephenson's (1992) *Snow Crash,* and, more particularly, the near-future megalopolises in which the technologies of human-machine interfaces are domesticated, such as the gigantic urban "Sprawl" of "BAMA...the Boston-Atlanta Metropolitan Axis" in William Gibson's (1984, p.57) *Neuromancer*, and the similarly immense but even more densely inhabited "Glop" of Marge Piercy's (1992) *He, She and It*. The imaginary town that provides the title of Ursula Le Guin's (1996) short story, "Ether, OR," serves as an ontologically ambiguous zone in which the ordinary is almost imperceptibly fused with the fabulous, as its middle-American inhabitants adapt to its shifting configurations and relocations ("People come through here say how can you stand living in a town that doesn't stay in the same place all the time, but have they been to Los Angeles? It's anywhere you want to say it is"; p.103). McHale (1987) notes that authors typically fashion such zones by "introducing an alien space *within* a familiar space, or *between* two adjacent areas of space where no such "between" exists'" (author's emphasis), and that even if the zone contains allusions to historical terrestrial spaces (Canada, occupied Germany, Oregon), it "in fact is located nowhere but in the written text itself" (pp.45–46). The "reality" of the zone, as Ballard (1984 [1964]) writes of Burroughs's novels, "is not some pallid reflection of a hypothetical external scene, its details and local color stitched into the narrative...but the self-created verbal reality of the next sentence and paragraph, like a track-laying train free to move about in all directions on a single set of rails" (p.106).

Thus, "teaching in the zone" signifies that curriculum inquiry, for me, is a textual practice shared with the many SF writers who, in Samuel Delany's (1988) words "posit a normal world—a recognizable future—and then an alternate space, sometimes largely mental, but always materially manifested, that sits beside the real world...conflicts that begin in ordinary space are resolved in this linguistically intensified paraspace" (p.31). I have included the parenthetic reference to Ballard's (1985 [1973]) *Crash* in my title because I admire it as one of the most intensely realized examples of the rhetorical heightening that can be achieved through such paraspatial textual strategies. Baudrillard (1991) describes *Crash* as the "first great novel of the universe of simulation, the world that we will be dealing with from now on" (p.319) and advances his own theorizing as a similarly evocative exercise:

> I am no longer in a state to "reflect" on something, I can only push hypotheses to their limits, snatch them from their critical zones of reference, take them beyond a point of no return. I also take theory into the hyper-space of simulation—in which it loses all objective validity, but perhaps it gains in coherence, that is, in a real affinity with the system that surrounds us. (Baudrillard, 1987, pp.36–37)

Baudrillard practices what Csicsery-Ronay (1991b) calls "the science fiction of theory" by inviting us to assume that "the system that surrounds us" *is* (has "real affinity with") a paraspatial zone. In this respect he implicitly accepts Ballard's (1985 [1974]) hypothesis that "we live in a world ruled by fictions of every kind" and, therefore, that "the most prudent and effective method of dealing with the world around us is to assume that it is a complete fiction....We live inside an enormous novel" (p.8).

As I have argued at length elsewhere (Gough, 1998b), I am convinced that we can initiate and sustain worthwhile inquiries in education by taking Ballard's proposition seriously. By this I do not necessarily mean that Ballard's assertion is a hypothesis to be tested "scientifically." Rather, I am suggesting that in practicing curriculum inquiry as a narrative art, I have sufficient reason to believe that Ballard's formulation refers to "real" conditions, with particular reference to the conditions under which I might need to judge my own and others' actions to be "prudent and effective" (or not). Rather than struggling with academic distinctions between stories of imagination and "reality," I choose to situate myself in the world-as-fiction as a "researcher-narrator" whose subjectivity is author of (that is, authorizes) his methodology.

Thus, in this chapter, I will elaborate and argue a case for adapting Ballard's and Baudrillard's assertions to science education. I will argue that the most prudent and effective method for teaching and learning science is to assume that the world in which science education is performed and represented is a fiction—a paraspatial, heterotopian zone; we should imagine that we are teaching and learning inside an enormous SF novel (or movie, or computer game, or…), that we are taking science curriculum into a hyperspace of simulation—metaphorically, into a *(Crash)* zone—in which we can push propositions and suppositions beyond their limits.

Science (Education) and SF: "The Boundary Between Science Fiction and Social Reality Is an Optical Illusion"

I have one further reason for using "the *(Crash)* zone" as a metaphor for the imagined space in which I prefer to teach science. *The Crash Zone* is the title of a thirteen-episode television series produced by The Australian Children's Television Foundation (ACTF)[2] in 1999. The series follows the adventures of five young people with a shared enthusiasm for computers, gaming, and the Internet who are hired by a small, struggling software company to test its computer games after school—"The Crash Zone" is their name for the space in which they work (they spraypaint the name on the door). I not only enjoyed viewing *The Crash Zone* with my then ten-year-old son, but also the ACTF invited me to write teaching activities for upper primary and lower secondary students to accompany the series. While the teaching activities I prepared (see, for example, Gough, 1999a, 1999b) were targeted to learning areas other than science education, writing them provided me with an opportunity to enact some of the orientations to curriculum work that I privilege here. One of these orientations, which I was pleased to see exemplified by *The Crash Zone* series itself, is expressed succinctly by Haraway's (1991) assertion that "the boundary between science fiction and social reality is an optical illusion" (p.149). *The Crash Zone* is in the genre of social realist comedy-drama—a junior high variant of "Friends"—and its only "science fictional" element is the inclusion of "Virgil," a virtual persona (an artificial intelligence program), as a character with whom the young people interact (a little like a Mario Brother that talks back incessantly). But Virgil is not depicted in a way that "objectively" separates him from the social reality of *The Crash Zone*—he is neither *deus ex machina* nor an object of "X-Files"-like investigation. Rather, he is seen as one among several actors—or as

Bruno Latour (1988, 1993) might say, "actants"—in a specific sociotechnological assemblage.

"A manifesto for cyborgs," in which Haraway's (1985) formulation of the illusory boundaries between SF and social reality first appeared, has been described as "groundbreaking" and "iconoclastic" (see, for example, Csicsery-Ronay, 1991b, p.393) because it presented and performed a new language for a socially democratic, materialist, and feminist politics. But Haraway's implicit pedagogical strategy is also continuous with, and complementary to, the call Marshall McLuhan (1960) issued some twenty-five years previously for "a basic strategy of culture for the classroom" (p.2) in which 'mass media' and 'mass entertainment' are "carefully set beside other kinds of art and narrative":

> It's misleading to suppose there's any basic difference between education and entertainment. This distinction merely relieves people of the responsibility of looking into the matter. It's like setting up a distinction between didactic and lyric poetry on the grounds that one teaches, the other pleases. However, it's always been true that whatever pleases teaches more effectively (p.3).

Similarly, "science" and "fiction" do not exist in separate domains but are culturally connected. This is not simply a matter of science and literature finding common meeting places in SF and other forms of popular media. Nor is it just a matter of scientific theories being translated into literary themes, a practice which long preceded the emergence of science fiction as a distinctive literary mode (for example, Copernican cosmology permeates the poetry of John Donne and concepts of disease formation are a distinctive feature of Emile Zola's novels). As Katherine Hayles (1984) demonstrates, "literature is as much an influence on scientific models as the models are on literature" (p.10), insofar as there is a two-way traffic in metaphors, analogies and images between them.

I have described elsewhere (Gough, 1991) some of the pleasures of learning with SF. My account is autobiographical and focuses on a small number of specific stories—by Arthur C. Clarke and Ursula Le Guin—that have significantly influenced my work in curriculum studies and teacher education. It was not written as an argument for using SF but, rather, as an example of its generative potential that might whet readers' appetites and motivate them to undertake their own inquiries. Nor is my account intended to demonstrate the "relevance" of SF to the subject matters and methods of conventional school science education. Like Nunan and Homer (1981), I do not advocate studying SF for the "textbook science" it may illustrate: "To do so would amount to little more than presenting the school-science orthodoxy in a slightly unorthodox way" (p.317). Rather, I believe that SF is a storytelling practice that science educators should welcome into the conceptual territory of science education with the deliberate intention that it should trouble the conventions and categories assumed by standard textbooks and ritual classroom activities.

The insights that can emerge from deliberately blurring distinctions between science and literature, and "fact" and fiction are convincingly demonstrated in *Primate Visions*, Haraway's critical history of the development and cultural effects of primatology. The introduction to *Primate Visions* is subtitled "the persistence of vision," and it is no coincidence that this is also the title of a story by John Varley (1978). Haraway (1989) writes:

> John Varley's science fiction short story, "The Persistence of Vision," is part of the inspiration for *Primate Visions*. In the story, Varley constructs a utopian community designed and built by the deaf-blind. He then explores these people's technologies and other mediations of communication and their relations to sighted children and visitors. The interrogation of the limits and violence of vision is part of the politics of learning to re-vision. (p.384)

Haraway exposes the "violence" that arises from the relationship between our vision—what, how, why, who, when, and where we choose to see—and those others (human, animal) who are the subjects and objects of (and who are subjected to and objectified by) our vision. In primatology, as in other disciplines, this violence is both literal and symbolic. As Haraway (1989) notes, "the commercial and scientific traffic in monkeys and apes is a traffic in meanings, as well as in animal lives" (p.1). Some of these meanings bear "the terrible marks of gender and race," because primatology has been a particularly important legitimating discipline for patriarchal, Eurocentric, and anthropocentric mythologies. Haraway is thus concerned to elucidate the ways in which the story telling practices of science, as exemplified by primatology, "structure scientific vision" and, in turn, construct myths of gender, race, and nature in our culture:

> Monkeys, apes, and human beings emerge in primatology inside elaborate narratives about origins, natures, and possibilities. Primatology is about the life history of a taxonomic order that includes people. Especially western people produce stories about primates while simultaneously telling stories about the relations of nature and culture, animal and human, body and mind, origin and

future. Indeed, from the start, in the mid-eighteenth century, the primate order has been built on tales about these dualisms and their scientific resolution (pp. 4–5).

Many of the "narratives about origins, natures, and possibilities" to which Haraway refers are sustained by popular media, such as the numerous film and video documentaries about Jane Goodall's work and various versions of the life and death of Dian Fossey (as in the movie, *Gorillas in the Mist)*. Other recent examples include novels such as William Boyd's (1991) *Brazzaville Beach*, Daniel Quinn's (1992) *Ishmael*, and Peter Høeg's (1996) *The Woman and the Ape*. SF has also produced many primate stories. For example, the Morlocks in H.G. Wells's (1895) *The Time Machine* are described as an "ape-like" evolutionary "degeneration" of humans and were inspired, in part, by a picture of a gorilla in an illustrated book of natural history which Wells read when he was seven (see Geduld, 1987, p.2). Wells (1934) himself describes *The Time Machine* as "a glimpse of the future that ran counter to the placid assumption of that time [the late Victorian era] that Evolution was a pro-human force making things better and better for mankind [sic]" (p.ix). In similar ways, monkeys and apes have inspired numerous images of human fears—including fears of what humans might become—in SF stories and movies such as the *Planet of the Apes* series.

The inspiration of Varley's SF story explicitly foreshadows one of the ways in which Haraway (1989) "reads" primatology, that is, "as science fiction, where possible worlds are constantly reinvented in the contest for very real, present worlds":

> I am interested in the narratives of scientific fact—those potent fictions of science—within a complex field indicated by the signifier SF....SF is a territory of contested cultural reproduction in high-technology worlds. Placing the narratives of scientific fact within the heterogeneous space of SF produces a transformed field. The transformed field sets-up resonances among all of its regions and components. No region or component is "reduced" to any other, but reading and writing practices respond to each other across a structured space. Speculative fiction has different tensions when its field also contains the inscription practices that constitute scientific fact. The sciences have complex histories in the constitution of imaginative worlds and of actual bodies in modern and postmodern "first world" cultures. (p.5)

I would describe the "transformed field" produced by "placing the narratives of scientific fact within the heterogeneous space of SF" as a *heterotopian zone* and, in later publications, Haraway (1994, 1997) coined the term "diffractions" as an optical metaphor for the "resonances" that may be set up within this textual space.

> My invented category of semantics, diffractions, takes advantage of the optical metaphors and instruments that are so common in Western philosophy and science. Reflexivity has been much recommended as a critical practice, but my suspicion is that reflexivity, like reflection, only displaces the same elsewhere, setting up the worries about copy and original and the search for the authentic and the really real.... What we need is to make a difference in material-semiotic apparatuses, to diffract the rays of technoscience so that we get more promising interference patterns on the recording films of our lives and bodies. Diffraction is an optical metaphor for the effort to make a difference in the world. (Haraway, 1997, p. 16)

By diffracting the primatology story through SF (and vice versa), Haraway also demonstrates the effectiveness of a McLuhanesque learning strategy: "Our time is a time for crossing barriers, for erasing old categories—for probing around. When two seemingly disparate elements are imaginatively poised, put in apposition in new and unique ways, startling discoveries often result" (McLuhan & Fiore, 1967, p. 10).

The results of adopting such a strategy are particularly apparent in the final chapter of *Primate Visions,* which alternates between "reading primatology as science fiction" and "reading science fiction as primatology." Haraway begins this chapter by using Isaac Asimov's (1964 [1952]) *The Second Foundation* to recapitulate the themes of *Primate Visions*. She then reviews the work of several women SF writers in the light of her reconstructed narratives of primatology. Haraway (1989) reasons that:

> Mixing, juxtaposing, and reversing reading conventions appropriate to each genre can yield fruitful ways of understanding the production of origin narratives in a society that privileges science and technology in its constructions of what may count as nature and for regulating the traffic between what it divides as nature and culture. (p. 370)

Primate Visions testifies to the potential effectiveness of SF in deconstructing, demystifying, and diffracting contemporary orthodoxies—in this case, the social, textual, and material history of primatology. Clearly, SF has mediated Haraway's own learning in important ways. The type of learning that Haraway models in *Primate Visions* is as applicable to school science education as it is to research in the history and philosophy of primatology. In my experience, school students require little encouragement to mix and juxtapose the narratives of "scientific fact" with the narratives of SF. Indeed, they may be more willing than their teachers to mix and juxtapose these "seemingly disparate elements" in critical and creative ways. The difficulty for science teachers is that many seem to have cast themselves in roles as "defenders of the faith"—defenders of the privileged status of science and technology—rather than "understanders" of the myths, narratives, and rituals which constitute science as a social and cultural practice in the contemporary world.

In pedagogical terms, Haraway is modeling a strategy for good teaching that Garth Boomer (1982) calls "connecting":

> The teacher is a senior reader of the school culture and special senior reader of the specialist subculture of the subject. Wittingly or unwittingly, he/she is demonstrating how to be a reader and maker of meaning....The more richly the teacher can spin a tapestry of metaphor and analogy into a "thick" redundant text of thinking about something new, the more likely it is that students will find a way in. If students are encouraged to spin out reciprocally their own webs of anecdote, metaphor and analogy, it is less likely that some will remain outside the next text. The art of generating apt analogy and metaphor is central to the 'reading' teacher's task. (pp.119–20)

In *Laboratories in Fiction: Science Education and Popular Media* (Gough, 1993b), I argued that it might be preferable for much of the routine work of science education (including much laboratory work) to be replaced by such "connecting" activities that would involve deconstructive readings of the cultural texts of scientific production, including primary sources (such as scientific reports), historical accounts of scientific work, the biographies and autobiographies of scientists, scientific journalism in print and electronic media, and images of science in the fine arts and popular media. In regard to the latter, I emphasized the significance of SF in its many and various forms as a "laboratory of ideas" in which meanings are subjected to experimentation (see also Gough, 1993a). More recently, I have offered another alternative that arises from rereading school laboratory work as a theater of material representations (Gough, 1998a). My purpose here is to demonstrate some ways of deconstructing and diffracting stories of science in their everyday manifestations. In the following section, I rehearse a reading of a recent news story that, following Haraway's example, deliberately mixes and juxtaposes the "reading conventions" of science and SF in such a way that neither is "reduced" to the other. My pedagogical strategy is to work to sustain a textual space in which diffractions of the narratives of science are enabled, that is, to deliberately position these stories in "the *(Crash)* zone" I evoked in the previous section.

Unleashing the Thylacine: Reading Science and SF in "the *(Crash)* Zone"

As Tyrone Slothrop wanders through the segmented Germany of Pynchon's (1973) *Gravity's Rainbow*, he observes: "here in the Zone categories have been blurred badly" (p.353). In a similar vein, I find it easy to think of teaching science in a zone in which the boundaries between science, SF, and social reality are illusory because I see that these categories have been blurred (though I would prefer to say *goodly* rather than "badly") within the discourses of our everyday social experience.

For example, "Unleashing the Thylacine" was headline news on the front page of Melbourne's daily broadsheet, *The Age,* on 13 May 1999. Under this headline, and dominating the upper-center section of the page, a large (16 x 27 cm) color photograph shows a tall cylindrical preserving jar holding a young specimen of what is identified as the now-extinct Tasmanian tiger.[3] Behind the upper-portion of the jar, an unidentified woman looks down at the slumped corpse, her face partly obscured and distorted by the curved glass. The photograph's caption states: "Back to the future: A perfectly preserved baby Tasmanian tiger offers the amazing prospect that the unique species could be rescued from beyond extinction."

The accompanying story, by reporter James Woodford (1999), includes the following passages:

> A twist of time may yet unleash the fabled Tasmanian tiger, assumed to be extinct for more than 60 years. Dr Mike Archer, the new director of the Australian Museum, says that it is time to bring the thylacine back to life, to reverse one of the great blots on the history of the colonisation of Australia. And scientists believe they may be able to clone a live thylacine—Jurassic Park-style—from the DNA taken from a perfectly preserved baby 'tiger' found in a jar in the museum. The tiger was plonked in the container in 1866, and by good luck was preserved in alcohol, rather than formalin, which would have destroyed its DNA. According to Dr Archer, it is feasible that people could own Tasmanian tigers as pets within 50 years. 'We've discovered the miracle bottle in which this time capsule is just waiting to pop back into life.' Dr Archer said cloning and genetic technology were moving so fast that within our lifetimes, one of the continent's unique creatures could be rescued from extinction. For Australians to turn their backs on this opportunity would be 'immoral'. His plan was supported by Dr Mike Westerman, senior lecturer in the genetics department at La Trobe University, who said that, with funds and application, it was possible Australia's largest carnivorous marsupial could be cloned in the 'not too-distant future'....The preserved pouch-young specimen was probably just old enough to be moving independently of its mother inside

the den, and has fur and the first signs of the species' unique stripes. Dr Archer said he was prepared to make it available to any scientist with a serious proposal to take the first steps to bring a thylacine back to life. 'Maybe in that preserved specimen, there's a way to get the DNA out. A researcher who came forward would be told instantly, "help yourself". There's probably a billion cells in that animal. That's a lot of cells to play with. 'There's a moral imperative here. It's not God's will that the thylacine went extinct. We did it.' (p.1)

There are no boundaries between science, SF, and social reality in this article. Both the journalist and the scientists deploy explicit and implicit references to well-known science fictions for their respective purposes. Not surprisingly, the journalist's appropriations—such as his references to the popular movies *Back to the Future* and *Jurassic Park*—are more obvious, but the scientists (assuming they have been quoted or paraphrased with reasonable accuracy) are complicit in authenticating the article's most persuasive message: that if scientists could possess an extinct organism's complete and perfectly preserved genome, its entire DNA code, then bringing a thylacine (or a dinosaur) back to life is merely a technical problem. Archer seems to suggest that if "there's a way to get the DNA out," if the "time capsule" can be opened, then the problem is almost solved, since the preserved cells are "just waiting to pop back into life." Both the journalist and the scientists reinforce the popular fiction that represents DNA as a list of instructions, analogous to a recipe in cookery or a computer program.

I doubt if either Archer or Westerman would defend this interpretation if pressed, since they are very unlikely to be ignorant of the flaws in the assumptions and reasoning they support with such enthusiasm here. Indeed, my ten-year-old son refuted their reasoning with very little prompting, since he had read (of his own choosing) Rob DeSalle and David Lindley's (1997) book, The Science of *Jurassic Park* and *The Lost World or, How to Build a Dinosaur* (the title of this book is a little misleading, because it actually is a rather smug litany of the science that is *absent* from Stephen Spielberg's movies and of how dinosaurs *cannot* be built *Jurassic Park*-style). My son thought that it might be possible to get the DNA out of the preserved thylacine, but he wondered where they would get the mother thylacine to put it in. As Jack Cohen and Ian Stewart (1994) point out, the belief that fossil DNA could be used to reconstruct a dinosaur (or, in this case, that embalmed DNA could be used to reconstruct a thylacine), is predicated on the view of DNA as "a genetic message transmitted from parent to offspring, a list of instructions like a glorified knitting pattern" (p.288). However, "the DNA message is not transmitted, but copied—subject to the complications of sexual reproduction—and the process whereby DNA code is turned into offspring also involves a parent" (p.290). Thus, dinosaur DNA "prescribes a dinosaur, within the context of another dinosaur" (p. 307). If the meaning of DNA is, like any other message, dependent on context, then hopes for "unleashing the thylacine" must presently be regarded as extremely slim.

I must emphasize here that I have no interest in arguing that Archer and Westerman are 'wrong' in any sense of the word, but I believe that as a science educator I should be able to deconstruct (that is, lay bare their construction of) the *Jurassic Park*-style spin they put on finding the preserved thylacine and that I should be able to facilitate learners' capacities to do likewise. Elucidating a less partial and less simplistic understanding of what scientists currently claim to 'know' about the functions and effects of DNA in reproductive and developmental biology is only one aspect of such a deconstructive analysis. I am ambivalent about using books like DeSalle and Lindley's (1997) as a resource for such a purpose, especially with children, because I do not want to undermine any child's belief in the *possibility* of restoring thylacines and dinosaurs to the world. In any case, I deplore the approach to using movies like *Jurassic Park,* exemplified by Dubeck, Moshier & Boss (1988) in *Science in Cinema: Teaching Science Fact through Science Fiction,* a book that is devoted to exposing scientific "misconceptions" and "pseudoscience" (p. x) in more than thirty science fiction movies. Their approach devalues the educative potential of SF by suggesting that it is deficient unless it illustrates the "one true story" of modern science "correctly" and also occludes the possibility of reading the films as critical and creative probes of issues in science, technology, and society that their makers and consumers consider to be problematic.

"Unleashing the thylacine" is the type of topical news story that students might reasonably be expected to use in both developing and demonstrating their "scientific literacy." For example, Australia's national curriculum profile for science includes a number of outcome statements at all levels of schooling that are concerned with evaluating the findings of scientific investigations, using science, and acting responsibly (see Australian Education Council, 1994). Table 1 displays a selection of these outcomes, together with examples of relevant issues that I suggest could be raised by the article.

TABLE: Selected outcome statements from science curriculum profile for Australian schools and related issues raised by "Unleashing the Thylacine"

Year levels	Selected outcome statement	Examples of issues related to "Unleashing the Thylacine"
3/4	Identifies ways science is used responsibly in the community.	Is trying to bring a thylacine—or a dinosaur—back to life a responsible use of science? What would you consider to be an irresponsible use of science?
5/6	Identifies the information needed to make decisions about an application of science	What further information would help you to decide whether cloning and DNA technologies should be used to bring an extinct species back to life? Should thylacines be cloned?
7/8	Identifies factors that influence people's perceptions of science	How do news items like this one (and movies like *Jurassic Park*) influence your and other people's perceptions of science?
9/10	Analyzes costs and benefits of alternative scientific choices about a community problem	What are some of the costs and benefits of using cloning and DNA technologies to bring extinct species back to life compared with other possible uses of these technologies (such as conserving presently endangered species)?
9/10	Analyzes the influence certain scientists have had on the ways we think about the world	What influence have the following scientists had (or might they potentially have) on the ways you and others think about the uses of DNA technologies: James Watson, Evelyn Fox Keller, Mike Archer, Ian Malcolm *(Jurassic Park*'s "chaotician")?
9/10	Analyzes the interactions between scientific developments and the beliefs and values of society	What social beliefs and values appear to be influenced by developments in biotechnology such as cloning and DNA sequencing?

I constructed Table 1 principally to demonstrate that the prospect of cloning the Australian Museum's thylacine pup could function as both a plausible and defensible focus for teaching and learning activities in school science education at several levels. However, I chose "Unleashing the Thylacine" as a focus for this essay because it is a ready-made interface with an imaginative 'zone' that cannot be reduced to the terms in which curriculum statements and conventional school science textbooks are framed. Although I am wary of essentializing or caricaturing my colleagues, I know that many science educators are likely to be tempted to categorize and draw boundaries between, the "scientific," "science fictional," and "socially realist" elements of this story—I found myself doing precisely that on first reading—but I will argue here that we should reject this "optical illusion" and deliberately cultivate the pleasures of their coexistence, including the diffractions produced by their juxtapositions.

Put crudely, the temptation for many science teachers is to encourage learners to "mine" articles like "Unleashing the Thylacine" for their "science fact" content—to proceed in the direction suggested by the title of Dubeck et al.'s (1988) book, cited above, of "teaching science fact through science fiction" and leave the zone of blurred boundaries behind them. When I brainstormed ideas for classroom activities that might be used to follow up on this news story with my first-year teacher education students, one of their most common suggestions was to "check out the Australian Museum's website" (some saw this as an activity for students, whereas others saw it as an aspect of teacher preparation). The Australian Museum's thylacine web pages are indeed a very useful resource for pursuing the types of outcomes and activities listed in Table 1, but they are explicitly located in the unambiguous space of "science fact" and cautious speculation. There are no references to *Jurassic Park*, no references to people owning Tasmanian tigers as pets within fifty years, and no open invitations to other biotechnology researchers to help themselves to the Museum's thylacine DNA.

One part of the site provides a brief natural history of the thylacine, its recent extinction, and its interrelationships with Tasmania's indigenous peoples. Another set of pages deals with fossil evidence of the thylacine and reports and rumors of recent sightings. A third series of pages presents "The cloning debate." The first of these pages, "Could the thylacine be cloned?,"[4] provides information about "three things that could be done in the foreseeable future with the thylacine genetic material" in much less optimistic terms than the Museum director's speculations. One of these options involves extracting the DNA in short pieces and putting it into bacteria to "make a genetic library." The DNA sequence "could be determined by techniques in current use," but "it would be very difficult to put the various pieces in the correct order." This is the only one of the three options that "can be done now," but this "is most unlikely ever to lead to anything remotely resembling a thylacine except in conjunction with the other approaches." The other two approaches "cannot be done yet" and "to be possible would require a lot more knowledge...and an investment in research into other species that would cost a great deal of money." In contrast to the director's generosity, this page asserts that should these new approaches be developed, they should "be proven on non-valuable specimens before using the Australian Museum pup." Rather than any researcher being "told instantly" to "help yourself," the page that deals with the question, "Should the thylacine be cloned?"[5] asserts that "thylacine specimens are so valuable that work should not be commenced without extensive consultation with both internal and external scientists and collection managers' and, further, suggests that "it could be argued...that the Museum's pup is so valuable as an object that it should not be damaged even to revive a species." The last page of the Museum's thylacine site invites visitors to cast their vote on the issue of whether or not the thylacine should be cloned[6].

It would be all too easy for teachers and students to restrict their examination of the plausibility and defensibility of cloning the thylacine to the types of "factual" information that can be obtained via the Australian Museum's website. But to do so would also make it far too easy to ignore many of the social implications of current trends in biotechnology that the "Unleashing the Thylacine" article invites us to consider. To analyze "the interactions between scientific developments and the beliefs and values of society" (see Table 1) demands that we also examine what Hayles (1990) calls "the assumptions that guide the constitution of knowledge in a given episteme" (p.xi). The juxtaposition of science, SF, and social reality that turns the prospect of resurrecting thylacines from extinction into front page news is "culturally conditioned, partaking of and rooted in assumptions that can be found at multiple sites throughout the culture" (Hayles, 1990, p.xi). For example, Hayles demonstrates that "different disciplines, sufficiently distant from one another so that direct influence seems unlikely...nevertheless focus on similar kinds of problems [at] about the same time and base their formulations on isomorphic assumptions" (p.xi). The operation of such isomorphic assumptions is implicit in Cohen and Stewart's (1994) assertion that "our current obsession with information technology and messages as bit strings has led us to focus almost exclusively on DNA as software and to ignore the contextual hardware in which it produces actions" (p.290). However, evidence for the existence of a cultural isomorphism between information understood as bits and bytes and DNA understood as software is not obvious when issues such as cloning a thylacine are presented in the careful languages of textbook science and the Australian Museum's website. But that evidence is clearly signposted in stories such as "Unleashing the Thylacine" where the extravagant claims of a media-savvy scientist pursuing his enlightened self-interests are juxtaposed with *Jurassic Park*. Neither Archer nor the journalist need to justify or explain the connection between extinct thylacine DNA and having a Tasmanian tiger "pop back into life" because, like the majority of the newspaper's readers, they tacitly acknowledge the cultural currency of *Jurassic Park's* central fiction. Mention dinosaurs and DNA in the same breath, and many of us will mentally replay the scene where the cartoon-animated "Mr DNA" explains: "a DNA strand, like me, is the blueprint for building a living thing, and sometimes animals that went extinct millions of years ago, like dinosaurs, left their blueprints behind for us to find. We just had to know where to look!"

I would not want students simply to compare Archer's exaggerations with the more restrained approach taken on the Museum's website and dismiss them as scientifically naïve or opportunistic attention-seeking. Archer's eminently quotable media bites are part of a smart sales pitch, exploiting the potential "domestication" of the research he promotes ("people could own Tasmanian tigers as pets within fifty years"), and invoking a "moral imperative" ("For Australians to turn their backs on this opportunity would be 'immoral'") with evangelical zeal ("It's not God's will that the thylacine went extinct"). Inviting students to consider the relationship between public interest and financial support for scientific research is an important aspect of analyzing "the interactions between scientific developments and the beliefs and values of society," but these are rarely dealt with in the idealized (and frequently heroic) accounts of scientific achievement recounted in textbooks.

I would also invite students to consider the wider ramifications of the biological determinism that is projected by the contiguity of Archer's assertions and the *Jurassic Park*-style message of DNA as the blueprint for life. I am reasonably confident that Archer would, if asked, explicitly reject an absolute genetic determinism—just like the scientists who speak and write about the Human Genome Project (which is mentioned as a significant resource in connection with one of the "things that can be done" with thylacine DNA on the Australian Museum's website). But, as

R.C. Lewontin (1994) concludes, these scientists "seem to be writing more to acknowledge theoretical possibilities than they are writing out of conviction" (p. 110). While mapping the human genome seems likely to be very useful in the diagnosis and treatment of illnesses with a unitary genetic cause, such as Huntington's chorea and cystic fibrosis, many scientists seem to expect much more of the project. For example, the molecular biologist Christopher Wills (1991) writes that "the outstanding problems in human biology...will *all* be illuminated" (p.2, my emphasis) by the Human Genome Project. Similarly, Cohen and Stewart (1994) quote from the program of the British Institute of Biology's 1993 symposium on Recent Advances in Human Genetics in the assertion that mapping the human genome will provide "the prime reference material for *all* biological and medical science" (p.463; emphasis in source). Cohen and Stewart (1994) comment: "Biologists will tell you that they don't say such naive things. They do" (p.463).

However, I suspect that such exaggerations are more a matter of cultural conditioning than individual naïveté. Lewontin (1994) summarizes some of the conditions under which the Human Genome Project promotes a deterministic ideology:

> The study of DNA is an industry with high visibility, a claim on the public purse, the legitimacy of a science, and the appeal that it will alleviate individual and social suffering. So its basic ontological claim, of the dominance of the Master Molecule over the body physical and the body politic, becomes part of the general consciousness. (p. 120)

Evelyn Fox Keller (1992) traces the circulation of this consciousness through the social milieu of the state, the universities, biotechnology corporations, and the media, producing an unquestioned consensus that "the model of cystic fibrosis is a model of the world" (p.290). In a disturbing example of how a medical model that begins with a genetic explanation for cystic fibrosis is transformed into an explanation of all social and individual variation, Keller (1992) quotes Daniel Koshland, then editor of *Science*, who was asked why the Human Genome Project funds should not be given instead to the homeless; he replied: "What these people don't realize is that the homeless are impaired...Indeed, no group will benefit more from the application of human genetics" (p.282).

The parallel point I want to make here is that analyzing "the interactions between scientific developments and the beliefs and values of society" in the context of "Unleashing the Thylacine" would entail encouraging students to consider how the allure of a *Jurassic Park*-style resurrection of the thylacine becomes not only irresistible to geneticists but also becomes part of a more general social consciousness. This means keeping students' attention focused on the cultural traces of such a consciousness and the diffractions that their juxtapositions with "scientific developments" produce. For example, on the day following the publication of "Unleashing the Thylacine," a cartoon by Michael Leunig on the editorial page of *The Age* (14 May 1999) whimsically alluded to other cultural images of the thylacine, such as their use in labeling and marketing a well-known brand of Tasmanian beer.

More significantly, one week later, a very prominently located, half-page feature article by Martin Flanagan (1999b), again in *The Age*, captured some of the social history of the thylacine since the European settlement of Tasmania. Titled "Tiger! Tiger! burning bright" the article begins: "Officially, the only Tasmanian tigers in existence are those in photographs. But...the animal lives on, burning into the collective consciousness a sense of irrecoverable loss, dread of an alien past and now, fantastically, hope of a comeback" (p.13). Flanagan notes that in addition to names such as tiger, hyena, dingo, native wolf, marsupial wolf, zebra wolf, and zebra opossum, in one location "the females of the species were called sluts" (at a time when "Aboriginal people were called crows"). He quotes a newspaper report from early this century that depicts the tiger as "Too Stupid to Tame," and the circumstances that led to a bounty being placed on the thylacine's head in 1888, which resulted in it being hunted to extinction. Flanagan's story concludes:

> It is now being suggested that science may be able to bring [the thylacine] back. Maybe, maybe not. What I do know is that when people meet a creature that is alien and strange and give it fearful names that enable them to strike it from their consciousness, what they find, with the passage of time, is what the sailor who shot the albatross in Coleridge's poem, *The Rhyme of the Ancient Mariner*, found. They have killed a part of themselves that they will henceforth seek in endless and fantastic ways. (p. 13)

When this article is juxtaposed with "Unleashing the Thylacine" we see another set of cultural imperatives being brought into play. Archer claims that "It's not God's will that the thylacine went extinct," and Flanagan brings on some literary heavyweights—William Blake and Samuel Coleridge—in solemn support. If, as Jonathon Porritt (1991) asserts, between 50 and 100 species of plants and animals become extinct each day, we might well ask what the thylacine did to get God, Blake, and Coleridge on its side? I am reminded of a *Monty Python* parody of a well-known hymn that (from memory) began as follows:

> All things dull and ugly/ All creatures short and squat/ All things rude and nasty/ The Lord God made the lot

Would similar media attention have been given to the Australian Museum finding some perfectly preserved spores of the fabled *Melanoswartus glabrocapitus* if it was assumed to have been extinct for more than sixty years? If "Unleashing the Blackhead Fungus" seems unlikely to be frontpage news, we might well ask what this means for analyzing "the interactions between scientific developments and the beliefs and values of society."

From where I stand, some of the most vivid diffractions of the thylacine story in the "zone" of blurred boundaries are enabled by Julia Leigh's (1999) novel, *The Hunter*, which at the time of writing had just been shortlisted for a major Australian literary award.[7] In Leigh's story, a man known only as M is hired by a biotechnology company to find and kill what may be the last thylacine and to retrieve its genetic material for the company's exclusive use. After several weeks alone in the Tasmanian wilderness, M wonders about the company's insistence on his mission being a one-man job when two might have been more efficient and considers the possibility that he is being set up for failure. M's thoughts, as Leigh (1999) imagines them, capture the motives behind the company's ruthlessness:

> Why one, why me? Was it possible the company no longer considered retrieval of genetic material a high priority?...He couldn't see why—what he had retrieved so far had earnt them, how much? Hundreds of millions, probably billions. The company needed him, in fact, was indebted to him. Who was of more value to a biotech company than a hunter: sampler and ensurer of exclusivity. Inbred thylacine, dodo, moa, mammoth, bunyip, yeti, girls with telekinetic powers, boys with an immunity to pain, the goose that laid the golden egg...mutations all, this was now the stuff that dreams—and wars— were made of. (p.50)

Passages like this are powerful reminders that any analysis of "the interactions between scientific developments and the beliefs and values of society" falls short if it rests with sociological abstractions and generalizations. Science and society are connected through embodied knowledges and expressed in stories that capture the particularities of existence, from fairy tales to *The X-Files*.

As a science educator, the questions that "Unleashing the Thylacine" raise for me do not really concern the plausibility or probability of the kind of genetic engineering imagined in *Jurassic Park* or the sinister operations of biotechnology corporations dramatized in *The Hunter*. Rather, the questions I would prefer to explore with learners are concerned with identifying the present circumstances and conditions—including any theoretic and technical knowledges that may be relevant and available—that make the events portrayed in these stories even *thinkable*. Who presently has power over and control of these circumstances, conditions, knowledges, and technologies? How is this power exercised? What safeguards exist against irresponsible uses of this power? What does it mean to be "responsible" or "irresponsible" in the circumstances and conditions made possible by these new knowledges and technologies?

Teaching science in "the *(Crash)* zone"—or *The Twilight Zone* if you prefer a less confronting metaphor—means enacting a pedagogy in which the truth claims of scientists are constantly rubbed up against the fictions that embody the relations of these claims with other cultural narratives that enrich and/or trouble their meanings. Relevant fictions are often suggested by science journalists, as in the suggestion that cloning a thylacine involves a *Jurassic Park*-style resurrection, but we should also be alert to other possibilities. With younger children, as the quote from *The Hunter* on page 266 suggests, we could equally well diffract "Unleashing the Thylacine" with the story of *Jack and the Beanstalk*.

Endnotes

1. References to the "the Zone" as a psychological space in which one's performance seems supernormal are also common in the literature of sport and physical adventure. For example, *In the Zone: Transcendent Experience in Sport* (Michael Murphy & Rhea White, 1995) documents numerous "moments of illumination, out-of-body experiences, altered perceptions of time and space, exceptional feats of strength and endurance [and] states of ecstasy" (p. 1) that have been reported by athletes and adventurers when they are enjoying a "peak performance" (p. ix, see also Andrew Cooper, 1998). The "zones" of sport and SF coincide in the movie *Field of Dreams*.

2. *The Crash Zone* was produced by the ACTF in association with The Disney Channel Australia and was first broadcast nationally by the free-to-air Seven Network from 13 February 1999. For further information, see http://www.actf.com.au

3. A photograph of the preserved pup can also be viewed on the Australian Museum's Website. See http://www.austmus.gov.au/thyline/09.htm
4. See http://www.austmus.gov.au/thlacine/09.htm
5. See "The cloning debate: Should the thylacine be cloned?" at http://www.austmus.gov.au/thylacine/12/htm
6. See "The cloning debate: Have your vote here!" at http://www.austmus.gov.au/thyline/vote.cfm
7. Flanagan (1999a) notes that there are two more thylacine novels on the horizon and that a recent article on the international biotechnology industry in *The London Review of Books* was illustrated with a photograph of a thylacine fetus.

References

Asimov, I. (1964 [1952]). *The second foundation.* New York: Avon.
Australian Education Council. (1994). *Science: A curriculum profile for Australian schools.* Carlton, Victoria: Curriculum Corporation.
Ballard, J.G. (1984 [1964]). Mythmaker of the 20th century. *Re/Search* (8/9), pp.105–107.
Ballard, J.G. (1985 [1973]). *Crash.* London: Paladin.
Ballard, J.G. (1985 [1974]). Introduction to the French edition of *Crash*, *Crash* (pp. 5–9). London: Paladin.
Baudrillard, J. (1987). The year 2000 has already happened. In Arthur Kroker & Marilouise Kroker (Eds.), *Body invaders: Panic sex in America* (pp.35–44). New York: St. Martin's.
Baudrillard, J. (1991). Ballard's *Crash. Science-Fiction Studies, 18* (3), pp.313–320.
Boomer, G. (1982). Ten strategies for good teaching. In Garth Boomer (Ed.), *Negotiating the curriculum* (pp.119–121). Gosford, NSW: Ashton Scholastic.
Boyd, W. (1991). *Brazzaville beach.* New York: William Morrow.
Bukatman, S. (1993). *Terminal identity: The virtual subject in postmodern science fiction.* Durham & London: Duke University.
Burroughs, W. S. (1964). *Nova express.* New York: Grove.
Burroughs, W. S. (1966 [1961]). *The soft machine.* (Revised ed.). New York: Grove.
Burroughs, W. S. (1967 [1962]). *The ticket that exploded.* (Revised ed.). New York: Grove.
Cohen, J., & Stewart, I. (1994). *The collapse of chaos: Discovering simplicity in a complex world.* New York: Viking Penguin.

Cooper, A. (1998). *Playing the zone: Exploring the spiritual dimensions of sports*. Boston: Shambhala.
Csicsery-Ronay, I. (1991a). Editorial introduction: Postmodernism's SF/SF's postmodernism. *Science-Fiction Studies, 18* (3), pp.305–308.
Csicsery-Ronay, Istvan. (1991b). The SF of theory: Baudrillard and Haraway. *Science-Fiction Studies, 18* (3), pp.387–404.
Delany, S. R. (1988). Is cyberpunk a good thing or a bad thing? *Mississippi Review* (47/48), 28–35.
DeSalle, R., & Lindley, D. (1997). *The science of Jurassic Park and The Lost World or, how to build a dinosaur*. London: HarperCollins.
Dubeck, L. W., Moshier, S. E., & Boss, J. E. (1988). *Science in cinema: Teaching science fact through science fiction films*. New York: Teachers College.
Fairclough, N. (1992). *Discourse and social change*. Cambridge, MA: Polity.
Flanagan, M. (1999a, 14 August). The hunt for Tasmania. *The Age Saturday Extra*, p.4.
Flanagan, M. (1999b, 20 May). Tiger! Tiger! burning bright. *The Age*, p.1.
Foucault, M. (1970 [1966]). *The order of things: An archaeology of the human sciences* (Alan Sheridan-Smith, Trans.). London: Tavistock Publications.
Geduld, H. M. (1987). *The definitive* Time Machine: *A critical edition of H.G. Wells's scientific romance*. Bloomington: Indiana University.
Gibson, W. (1984). *Neuromancer*. New York: Ace.
Gough, N. (1991). An accidental astronaut: learning with science fiction. In George Willis & William H. Schubert (Eds.), *Reflections from the heart of educational inquiry: Understanding curriculum and teaching through the arts* (pp.312–320). Albany: State University of New York Press.
Gough, N. (1993a). Environmental education, narrative complexity and postmodern science/fiction. *International Journal of Science Education, 15* (5), pp.607–625.
Gough, N. (1993b). *Laboratories in fiction: Science education and popular media*. Geelong: Deakin University.
Gough, N. (1998a). 'If this were played upon a stage': School laboratory work as a theatre of representation. In Jerry Wellington (Ed.), *Practical work in school science: Which way now?* (pp.69–89). London: Routledge.

Gough, N. (1998b). Reflections and diffractions: Functions of fiction in curriculum inquiry. In William F. Pinar (Ed.), *Curriculum: Toward New Identities* (pp. 94–127). New York: Garland.

Gough, N. (1999a). The lost worlds of *The Crash Zone*, The Australian Children's Television Foundation. http://www.actf.com.au.

Gough, N. (1999b). Where in the world is...cyberspace?, The Australian Children's Television Foundation. http://www.actf.com.au.

Haraway, D. J. (1985). A manifesto for cyborgs: Science, technology and socialist feminism in the 1980s. *Socialist Review, 15* (2), pp.65–107.

Haraway, D. J. (1989). *Primate visions: Gender, race, and nature in the world of modern science*. New York: Routledge.

Haraway, D. J. (1991). *Simians, cyborgs, and women: The reinvention of nature*. New York: Routledge.

Haraway, D. J. (1994). A game of cat's cradle: science studies, feminist theory, cultural studies. *Configurations: A Journal of Literature, Science, and Technology, 2* (1), pp.59–71.

Haraway, D. J. (1997). *Modest_Witness @Second_Millennium. FemaleMan©_Meets_OncoMouse™: Feminism and technoscience.* New York and London: Routledge.

Hayles, N. K. (1984). *The cosmic web: Scientific field models and literary strategies in the twentieth century*. Ithaca, NY: Cornell University.

Hayles, N. K. (1990). *Chaos bound: Orderly disorder in contemporary literature and science*. Ithaca, NY: Cornell University.

Høeg, P. (1996). *The woman and the ape* (Barbara Haveland, Trans.). London: The Harvill.

Keller, E. F.. (1992). Nature, nurture, and the Human Genome Project. In Daniel J. Kevles & Leroy Hood (Eds.), *The code of codes: Scientific and social issues in the Human Genome Project* (pp.281–299). Cambridge, MA: Harvard University.

Latour, B. (1988). *The Pasteurization of France* (Alan Sheridan & John Law, Trans.). Cambridge, Ma: Harvard University.

Latour, B. (1993). *We have never been modern* (Catherine Porter, Trans.). Cambridge, MA: Harvard University.

Le Guin, U. K. (1996). *Ether, or, unlocking the air and other stories* (pp.95–123). New York: HarperCollins.

Leigh, J. (1999). *The hunter*. Ringwood, Victoria: Penguin.

Lessing, D. (1980). *The marriages between zones three, four and five*. London: Jonathan Cape.

Lewontin, R.C. (1994). The dream of the human genome. In Gretchen Bender & Timothy Druckrey (Eds.), *Culture on the brink: Ideologies of technology* (pp.106–127). Seattle: Bay Press.
McHale, B. (1987). *Postmodernist fiction*. New York and London: Methuen.
McHale, B. (1992). *Constructing postmodernism*. London and New York: Routledge.
McLuhan, M. (1960). Classroom without walls. In Edmund Carpenter & Marshall McLuhan (Eds.), *Explorations in communication* (pp.1–3). Boston: Beacon.
McLuhan, M., & Fiore, Q. (1967). *The Medium is the Message*. New York: Bantam Books.
Murphy, M., & White R. A. (1995). *In the zone: Transcendent experience in sport*. New York: Penguin.
Nunan, E. E., & Homer, D. (1981). Science, science fiction, and a radical science education. *Science-Fiction Studies* (8), pp.311–330.
Piercy, M. (1992). *He, she and it*. New York: Alfred A. Knopf.
Porritt, J. (1991). *Save the earth*. North Ryde, NSW: Angus & Robertson.
Pynchon, T. (1973). *Gravity's Rainbow*. New York: Viking.
Quinn, D. (1992). *Ishmael*. New York: Bantam Books.
Stephenson, N. (1992). *Snow crash*. New York: Bantam Books.
Strugatsky, A., & Strugatsky, B. (1978). *Roadside Picnic* (Antonina W. Bouis, Trans.). London: Victor Gollancz.
Varley, J. (1978). *The persistence of vision*. New York: Dell.
Wells, H.G. (1895). *The time machine*. London: Heinemann.
Wells, H.G. (1934). *Seven famous novels*. New York: Knopf.
Wills, C. (1991). *Exons, introns, and talking genes: The science behind the Human Genome Project*. New York: Basic Books.
Woodford, J. (1999, 13 May). Unleashing the thylacine. *The Age*, p.1.

Chapter Eleven
Pedagogies of Science (In)formed by Global Perspectives: Encouraging Strong Objectivity in Classrooms
Annette Gough

Those of us with a science degree of some sort have grown up with–and have often been enculturated by—the notion that the "nature of [modern] science" can be characterized by its claims to objectivity, rationality, and truth. Some have even engaged in denying any bias or political agenda in the production of scientific knowledge, exemplifying Marion Namenwirth's (1986) statement that "Scientists firmly believe that as long as they are not conscious of any bias or political agenda, they are neutral and objective, when in fact they are only unconscious" (p. 29). Sandra Harding (1994) provides an explanation for this absence of "science criticism" in scientific communities:

> Failing to locate any significant critical studies of the sciences in universities, and especially in science departments, indicates to students that no one thinks these studies important for learning to do science or for making reasoned decisions about scientific issues in public life. This is unfortunate, since, as the postcolonial accounts show, philosophical, sociological, and historical assumptions form part of scientific understanding *about nature.* Scientists unknowingly use distorting cultural assumptions as part of the *evidence* for their research results if they are taught that social studies of science are irrelevant to doing science, and that they should assiduously "avoid politics" rather than learning how to identify cultural features in their scientific assumptions, and how to sort the distorting and knowledge-limiting from the knowledge-enlarging cultural values and interests (p.329).

This chapter is concerned with the apparent lack of awareness/ consciousness among science educators (as well as scientists) of the global culture of which science is a part and discusses the need to change both the knowledge and classroom practices of science educators, particularly through adopting positions such as Sandra Harding's "strong objectivity." One of Harding's strengths is that she does not throw out the baby with the bath water and totally reject Western science. Rather, she recognizes "some of the still immensely valuable elements of the European philosophic tradition" and argues for retrieving and transforming them "into tools useful in today's multicultural and postcolonial world" (1998, p.x). Hence, rather than abandoning objectivity, Harding (1993c) believes that "the notion of objectivity is useful in providing a way to think about the gap that should exist between how any individual or group wants the world to be and how in fact it is" (p.72). She thus argues for a strong objectivity which "requires that the subject of knowledge be placed on the same critical, causal plane as the objects of knowledge" (p.69) by adopting feminist and postcolonial standpoints which she sees as "also firmly within conventional dynamics of northern sciences committed to critical self-evaluation" (Harding, 1998, p.22). Harding's notion of strong objectivity is similar to Haraway's (1991) argument for "a doctrine and practice of objectivity that privileges contestation, deconstruction, passionate construction, webbed connections, and hope for transformation of systems of knowledge and ways of seeing" (pp.191–192).

How Harding characterizes standpoint approaches can be explained by contrasting them with empiricist approaches (see A. Gough, 1994). For example, whereas empiricism tries to fit projects into prevailing standards of "good science" or "good philosophy" and just tell a different "one true story" about reality in the Enlightenment tradition, Harding (1993) argues that feminist standpoint theory

> sets out on a rigorous "logic of discovery" intended to maximise the objectivity of the results of research and thereby to produce knowledge that can be *for* marginalized people (and those who would know what the marginalized can know) rather than *for* the use only of dominant groups in their projects of administering and managing the lives of marginalized people (p.56).

This journey leads to stronger objectivity and less partial stories. The two approaches can be contrasted by looking at their differing stances on the subject or agent of knowledge (see Table on p.275). In criticizing current practices and suggesting changes, my aim is not to argue *against* science education but, rather, to argue *for* a more democratic science education, one that will relate a less partial, less distorted view of the natural world through science studies and one which will better relate to students' experiences and interests. I share Harding's (1998) commitment to "strengthening the objectivity of understandings of modern sciences and technologies, of the sciences and technologies of other cultures, and of historical and possible future relations between them" (p.18), within my goal of a (post) modern science education.

Table: Comparison of the approaches to the subjects/agents of knowledge and to knowledge production in empiricist epistemology and standpoint epistemology (from A. Gough, 1994, after Harding, 1993, pp. 63–65)

Empiricist Epistemology	Standpoint Epistemology
•subject is culturally and historically disembodied or invisible because knowledge is by definition universal	• subject is embodied and visible because the lives from which thought has started are always present and visible in the results of that thought
• the subject is different in kind from the objects whose properties scientific knowledge describes and explains, because the latter are determinate in space and time	• because the subjects of knowledge are embodied and socially located, they are not fundamentally different from objects of knowledge
• knowledge is initially produced by individuals and groups of individuals not by culturally specific societies or subgroups in a society	• communities and not primarily individuals produce knowledge
• subject is homogeneous and unitary because knowledge must be consistent and coherent	• subjects/agents of knowledge are multiple, heterogeneous, and contradictory or incoherent

Beyond a Universal Science (Education)

In the world view held by most science educators, science (i.e., modern Western[1] science) is seen as universal—applying to everyone and everything—and this view is not questioned. This belief in the universality of knowledge is based on at least two assumptions: "Firstly, that there is one uniquely correct ordering of the phenomena of the natural world. Secondly, that there is a set of procedures sufficiently powerful to determine what that ordering is" (Turnbull 1991, p.9). Increasingly, during the past few decades, this view of how scientific knowledge is produced has been questioned from a number of different perspectives, including feminist, postcolonialist, sociological, anthropological, and critical and cultural studies.[2]

Central to these critiques is a concern with knowledge production in science and with asking questions such as, "who can be a 'knower,' what can be known, what constitutes and validates knowledge, and what the relationship is or should be between knowing and being" (Stanley & Wise, 1990, p.26). Thus there is a skepticism "about the possibility of a general or universal account of the nature and limits of knowledge, an account that ignores the social context and status of knowers" (Alcoff & Potter, 1993, p.1). A concern with knowledge production is also consistent with the view of pedagogy proposed by Lusted (1986, in Gore 1993) who argues that pedagogy, as a concept, "draws attention to the *process* through which knowledge is produced. Pedagogy addresses the 'how' questions involved not only in the transmission or reproduction of knowledge but also in its production" (p.4). It includes both instruction and social vision (Gore, 1993) and, in the case of science education, includes content, process, and discourse (Barton, 1997). This view of pedagogy is different from that usually practiced in science classrooms, where the focus tends to be on the transmission and reproduction of knowledge. Practicing a pedagogy which focuses on the process of knowledge production means acknowledging the politics of the production processes and the political and cultural contexts in which they are situated. Such a view is consistent with enacting "strong objectivity" in science.

My own journey into this "new" territory started through two parallel yet eventually converging concerns. The first was an interest in feminist critiques of science and the implications of these for science education theory and practice (see A. Gough, 1998). The second came through my involvement in environmental education and interest in the relationship between science and the environment (see A. Gough, 1997). Guided by many authors, but particularly by Sandra Harding and Carolyn Merchant, my focus soon changed from just feminist critiques of science to reading and reflecting upon cultural, postcolonialist, sociological, and other critiques and relating these to science education (see A. Gough, 1998). I have thus come to believe that we need to recognize the socially constructed, gendered, and multicultural nature of science in its global context within science education pedagogies and to work toward a more democratic concept of science in our science teaching practices, and such a belief frames this chapter.

The critics of modern science as it is currently practiced are generally not "antiscience" but, rather, seeking a *better science* for a variety of political aims. Some would like to see scientists being more self-critical about the political and social origins of their research, "pointing out how better understandings of nature result when scientific projects are linked with and incorporate projects of advancing democracy; [and that] politically regressive societies are likely to produce partial and distorted accounts of the natural and social world" (Harding, 1993a, p.ix). Others want to see science redeem its tarnished ideals from internal abuse and external impurities. A third group would like to create new scientific methods grounded in the social needs of communities accountable to the people whose lives are invaded by the effects of scientific super-industrialization. My interest is in the implications of these critiques for how science education is practiced in schools—with particular reference to those critiques concerned with developing a democratic approach to science and providing opportunities for the public to become actively involved in aspects of society that impinge upon the quality of their lives.

In the following sections I discuss some of the challenges to science education pedagogy that are posed by a recognition of the socially constructed nature of science from feminist, multicultural, and global perspectives. I then discuss some of the models being proposed for changing science education practices, and conclude with a discussion of the implications of Harding's "strong objectivity," and other pedagogies of cultural studies of science, for classroom practices consistent with a (post) modern science (education).

Is Science Gendered?

The association of woman and nature as suppressed binary oppositions to "man" and culture has been a focus for research in gender and science, particularly as this binary opposition is reflected in the language of science (see, for example, Harding, 1991, 1998; Keller & Longino, 1996; Merchant, 1980).

As Evelyn Fox Keller (1985) argues,

> science is the name we give to a set of practices and a body of knowledge delineated by a community, not simply defined by the exigencies of logical proof and experimental verification. Similarly, masculine and feminine are categories defined by a culture, not by biological necessity (p.4).

However, Ruth Hubbard (1981) notes that, "while many people have long acknowledged that social context inevitably affects and often structures perceptions of reality, most continue to believe that the scientific methodology guarantees objectivity to scientists and science" (p. 213). More recently, Ruth Berman (1989) has argued that this belief is beginning to lose ground:

> The vision of science, rising with magisterial authority above the political battles raging below, has grown somewhat dim for women and for some men. The impartiality of its pronouncements has been questioned on several counts by feminists and others, its claim to objectivity described as myth. It is now being seen as a potent agent for maintaining current power relationships and women's subordination (p.224).

A strong theme throughout the feminist critiques of science is concern with criticizing the objective, value-free claims made of science. As Hubbard (1981) argues, "An era's science is part of its politics, economics and sociology: it is generated by them and in turn helps to generate them" (p.218). Bleier (1986) similarly argues that "scientists cannot simply hang their subjectivities up on a hook outside the laboratory door...as is the case for everyone else, scientists bring their beliefs, values, and world views to their work" (p.63), and these affect their research questions, assumptions, interpretations, and activities. Hubbard (1981) also argues that to become conscious of the biases introduced by our implicit, unstated, and often unconscious beliefs about the nature of reality is more difficult than anything else we do, but "we must try to do it if our picture of the world is to be more than a reflection of various aspects of ourselves and of our social arrangements" (pp.218–219). Harding (1986) further develops this notion by arguing that "objectivity never has been and could not be increased by value-neutrality. Instead, it is commitments to antiauthoritarian, anti-elitist, participatory, and emancipatory values and projects that increase the objectivity of science" (p.27). Much more has been written on the subjectivity/objectivity dialectic, and it is not within the scope of this chapter to resolve the issue. However, in later sections I discuss the implications of strong objectivity for science education practices.

One of the puzzles in the area of gender and science education research is that, while all of the researchers are concerned about the absence of women in science, few are willing to entertain the notion that science is a masculine discourse. Solomon (1997), for example, states that radical feminist critiques of science and society—she cites Harding (1991) as an example—"is uncomfortable for science educators," as they take modern science as a given rather than seeking "to ensure that a new science emerges" (p.409) through the STS (Science Technology and Society) movement. However, while many may, not all science educators agree with Solomon. Brickhouse (1994), who was a colleague of Harding's at Delaware, argues that science needs to be taught so that students understand its multicultural nature, its controversial nature, and its relationship to the world so that students' masculine image of science will be changed. Kyle (1999) writes of multiple knowledges, rather than a singular or universal knowledge, and the need for science education to be dynamic, linked to life world experiences and how students can take action in the world to create the future. Baker (1998) suggests strategies for overcoming the conception of science as a white European as well as male domain, and Osborne and Barton (1998) are developing a liberatory science pedagogy grounded in feminist theory which will help science educators "to critically examine from multiple perspectives the intermingling of science, power and status in science class, and to envision a 'socially-transformative' science education" (p.252).

Both Kreinberg and Lewis (1996) and Rosser (1997) have adapted Schuster and Van Dyne's (1984) liberal arts model of curriculum transformation for science education. Their versions of this (rather too linear) model has six stages:

Stage 1: Absence of women in science not noticed.

Stage 2: The search for the missing women in science.

Stage 3: Why are there so few women in science?

Stage 4: Studying women's experience in science.

Stage 5: Challenging the paradigm of what science is.

Stage 6: The transformed, gender balanced (gender-free) science curriculum.

Of particular relevance to curriculum transformation are stages 5 and 6. Stage 5 is concerned with "challenging the paradigm of what science is" (Kreinberg & Lewis, 1996, p.195) through focusing on "science done by feminists/women" and thus "improving the quality of science" (Rosser, 1997, p.4). Their respective conceptions of science in Stage 6 appear to differ. In Stage 6 science is "redefined and reconstructed to include us all" (Rosser, 1997, p.4) in "the transformed, reconstructed gender-free [science] curriculum" (Kreinberg & Lewis, 1996, p.197). Krienberg and Lewis's introduction of a "gender-free" curriculum seems to be a return to the objectivist model of modern science, rather than, recognizing the socially constructed nature of both gender and science and continuing this argument. While Rosser's reconstruction of science education to include "us all" is potentially a little more ambiguous, it can also be read as a return to a universal science for a universal subject. Neither conception seems to be consistent with Harding's strong objectivity and the generation of less partial, less distorted stories (see Table 1). On this basis, perhaps Solomon is right, and science educators *are* uncomfortable with having their modern science destabilized. However, this does not mean that we should abandon attempts to do so. More recently, Solomon (1999) has continued her defence of modernist science against postmodernist criticism, but she has made a few concessions. She recognizes that students reject rigorous science education's insistence "on a complete abjuration of all that is not logical" (1999, p.7) because they want to claim more varied attitudes and accept that there may be a need for a new science curriculum that "will aim to show young people a science which is lighter on logic and abstraction, stronger on involvement and active evaluation, and intimately woven into the aspirations and concerns of citizens" (1999, p.9). Other aspects of Solomon's (1999) proposal are discussed later in this chapter.

Just as science education curriculum clings to a modernist science position, so does much research in gender and science education. With few exceptions, such research tends to adopt what feminist researchers see as masculinist (empiricist) paradigms of educational research. In particular, much gender and science education research tends merely to report quantitative findings from isolated studies. As Krockover and Shepardson (1995) assert,

> Fuller, richer images of the multiple contexts and identities are clearly afforded by utilizing the full array of qualitative methodologies….Studies need to move from short-term analysis to longitudinal investigations….Finally, researchers ought to establish and develop collaborative partnerships with practitioners (p. 223).

Each of these different approaches noted by Krockover and Shepardson, and more, have been debated and used by feminist researchers in other fields for some time. But science education researchers concerned with gender issues have tended to be preoccupied with providing statistics rather than rich images. This situation is gradually changing. Baker and Leary (1995) also comment on the previous preoccupation with quantitative analysis and choose to pursue one of the emerging emphases in nonpositivist research in this area—interview and other approaches grounded in the work of Carol Gilligan (1982)—and Joan

Solomon (1997) reports on a three year GIST (Girls into Science and Technology) Project which uses action research methodologies.

There are increasing dilemmas for science education researchers and pedagogues interested in gender issues. These come from a number of sources:

- encouragement by feminist science philosophers such as Sandra Harding (1991, 1998) who have destabilized previously held 'truths' about the nature of science,
- the continuing debate about feminist research methodologies from researchers such as Patti Lather (1991) and others,
- the importance of language and discourse in science (see, for example, Haraway, 1991; Keller, 1992) and other insights from (particularly feminist) poststructuralism (see, for example, Weedon 1987, 1999), and
- the increasing acceptance of multiple subjectivities (gender, race, ethnicity, class) in the (de)construction of meanings and identities.

Some of the challenges arising from these dilemmas are taken up later in this chapter.

Is Science Multicultural?

The notion of multiculturalism in relation to science is relatively new and "the term 'multicultural' is notoriously vague: it sweeps under its blanket generality a tangle of confusions and uncertainties about how cultures can or should relate to each other, and how their world-views relate to the world" (Kuriyama, 1994, p.337). Sandra Harding (1994) uses the term "multicultural" as a counter to "universal science," i.e. science that is Eurocentric or Western or Northern, depending on how one looks at it, and distinguishes it from postcolonial science studies. Instead of "multicultural," Kuriyama (1994) proposes "comparability," because "it evokes in its very etymology the primacy of parity, the quest for a way of imagining cultures that assert equality while recognizing difference" (p.338). While this alternative has its attractions, for the purposes of this chapter, and despite the term being vague, "multicultural" will be used for its flexibility.

The notion of multiculturalism challenges the traditional stance of scientific knowledge being context- and value-free (objective), i.e., it challenges the universal science view "that modern sciences are uniquely successful exactly because they have eliminated cultural fingerprints from their results of research" (Harding 1994, p.304). Harding's arguments for science being multicultural can be summarized as:

> • what is called Western science incorporates significant elements from non-Western cultures;
> • there are knowledge traditions outside the West that have legitimate claims to be considered sciences;
> • Western science, no less than knowledge elsewhere, is enmeshed in a dense network of broader cultural values and assumptions.

These broader values and assumptions include gender, race, class, religion, among others.

Harding answers her question, "Is science multicultural?" in the affirmative, and her respondents (Cohen, 1994; Farquhar, 1994; Kuriyama, 1994) in the *Configurations* set of papers, concur, albeit to varying degrees.

With struggles over the social standing of science and declining enrollments in senior secondary school science subjects and university science courses, it is timely to re-think the content as well as the pedagogy of science education in schools. As Harding (1993b) argues,

> Science education has suffered from its lack of attention to such phenomena as the "racial" economy of science, perhaps fearing that any but the most minimal admission of error in the past...would generate the accurate perception that not enough has been done to block such antidemocratic and ignorance producing tendencies today (p.6).

Our thinking of science teaching and learning from a multicultural perspective involves consideration of a number of proposals. Harding (1994, pp.322–326) discusses the alternatives envisioned in Third World postcolonial science analyses to the continued suppression of indigenous scientific goals, practices, and culture by Western ones. These are:

> • integrate endangered Third World sciences into modern sciences;
> • strengthen indigenous scientific traditions through adopting those parts of modern science that could be integrated into *them*;
> • "de-link" Third World scientific projects from Western ones;

- accept that Third World sciences and their cultures can provide useful models for global sciences of the future. She also proposes three contributions to viable future sciences that can be made from a Western perspective (Harding, 1994):
- relocate the projects of sciences and science studies that originate in the West on the more accurate historical map created by the new postcolonial studies, i.e., a "new 'science education' about the history of scientific traditions" (p.327).
- add a new education *in* the sciences both for schoolrooms and for public discussion in the media through which a citizenry educates itself, i.e., "more accurate and critical studies of sciences in their historical context should form an important part of science education as well as general education" (p.329).
- consider the above and "other such distinctive tasks as a progressive project for constructing fully modern sciences that creatively develop key elements of the Western cultural legacy" (p.330) and postmodernize them.

While many science educators (and scientists) may not feel comfortable with these proposals they do have some common ground with the view of border (antiracist) pedagogy (of postmodern resistance) expounded by Henry Giroux (1991). He calls for "a notion of border pedagogy that provides educators with the opportunity to rethink the relations between the centers and margins of power" (p.247) which should be implemented by:

- offering students the opportunity to address texts as social and historical constructions and educating them "to perform ideological surgery on master-narratives based on white patriarchal, and class-specific interests" (p.249);
- engaging students in understanding "how racist practices develop, where they come from, how they are sustained, how they affect dominant and sub-ordinate groups, and how they can be challenged...[as] a discourse about economics, culture, politics, and power" (p.250);
- offering students the opportunity "to air their feelings about race from the perspective of the subject positions they experience as constitutive of their own identities" (p.250);
- educators understanding "how the experience of marginality at the level of everyday life lends itself to forms of oppositional and transformative consciousness" (p.251);

- analyzing racism "not only as a structural and ideological force, but also in the diverse and historically specific ways in which it emerges" and disavowing "all claims to scientific method or for that matter to any objective or transhistorical claims" (p.252);
- redefining "how the circuits of power move in a dialectical fashion among various sites of cultural production" (p.253);
- educators adopting "a pedagogy that provides a more dialectical understanding of their own politics and values...break[ing] down pedagogical boundaries that silence them in the name of methodological rigor or pedagogical absolutes...develop[ing] power-sensitive discourse...so that their classrooms can engage rather than block out the multiple positions and experiences" (p.254).

In his editorial introduction to a recent issue of *Journal of Research in Science Teaching*, William Kyle (1999) raises some of these issues within a context of science education: "The totality of an education in science is equally as much oriented toward social justice, critical democracy, empowerment, action-taking, investing in our future's intellectual capacity as it is about constructing conceptual understanding of the world" (p.255). It is the literature that is moving in this direction which I discuss in the next section.

Shifting Relations in Current Science Education Practices
Historically, over the past century, there has been a long process of change in which science has come to take its place alongside, and perhaps even ahead of, the humanities as a central feature of the curriculum code, often described as the "competitive academic curriculum," which sorts students destined for tertiary studies and careers in the professions. It was not always so. In the first part of this century, science was generally perceived as lacking the intellectual demands of the classics and was often associated with the trades. As science has acquired respectability, so what counts as science has in turn been constrained. As part of the social process of forming a "science curriculum," university research science has become the underlying model upon which science has been created. Applied science has been pushed to the margins, analysis has come to replace genuine experimentation. School science (in the Western and English speaking worlds at least) has kept its distance from industry, from medicine, and from everyday applications (Medway, 1993; Reiss, Millar, & Osborne,

1999; Solomon, 1999). Contemporaneously (and perhaps not coincidentally) with this development, there has been a declining interest in science by young people and declining enrollments in (particularly physical) sciences at upper-secondary and university levels (Dekker & De Laeter, 1997; Driver et al., 1996; Solomon, 1999).

There have been decades of attempts by curriculum developers and others to create curricula that take science out of the research laboratory; however, these have mostly failed, except in courses that are of relatively low status or marginal to the mainstream career routes to the professions. The tradition of science as invention, which Medway (1993) identifies with Edison, appears to have no home in the school science laboratory, except in the case of a few enthusiasts. Within all of this there is a kind of scientific illiteracy, as the science education that is being made available is not focused upon developing an understanding of science as a fully social process. As Sandra Harding (1993) argues,

> There are few aspects of the "best" science educations that enable anyone to grasp how nature-as-an-object-of-knowledge is always cultural: "In science, just as in art and life, only that which is true to culture is true to nature.".These elite science educations rarely expose students to systematic analyses of the social origins, traditions, meanings, practices, institutions, technologies, uses, and consequences of the natural sciences that ensure the fully historical character of the results of scientific research (p.1).

However, as part of the millennium fever ("Beyond 2000"), and for other reasons—including the ever-present international competitiveness about results from testing student achievement (such as the Third International Mathematics and Science Study), a growing acceptance of the need to address gender issues in science education, concerns about declining student interest in science, calls for great public scientific literacy—there are currently developments in several countries which are concerned with re-vitalizing science education. But are these developments addressing the deficiencies identified by Harding; are they consistent with a society that is postmodern, and how will science educators change their knowledge and practices?

Carolyn Merchant (1980) writes of another period of change—in the seventeenth century—when the mechanical view of nature characteristic of modern science replaced other ways of thinking:

> The mechanical view of nature now taught in most Western schools is accepted without question as our everyday, common sense reality—matter is made up of

atoms, colors obey the law of inertia, and the sun is the center of our solar system. None of this was common sense to our seventeenth-century counterparts. The replacement of the older, "natural" ways of thinking by a new "unnatural" form of life-seeing, thinking, and behaving did not occur without struggle. The submergence of the organism by the machine engaged the best minds of the times during a period fraught with anxiety, confusion, and instability in both the intellectual and social spheres (p.193).

The parallels with the present day and the "postmodern turn" are, for some, equally "a period fraught with anxiety, confusion, and instability in both the intellectual and social spheres"—yet neither the past nor the present periods of instability are discussed in school science classrooms, and the future, if discussed at all, is usually discussed in terms of technological change. For example, Solomon (1999) states that the value of science education "now relies on the social salience of the issues of new technology in our culture to which it would be linked" (p.14).

This is certainly the case with the Nuffield Foundation's proposed structure for the school science curriculum in England and Wales, "Beyond 2000" (Reiss, Millar, & Osborne, 1999). The authors recognize that the present curriculum overemphasizes content isolated from meaningful contexts, is too oriented toward sorting students through formal examinations, and is of declining interest to students. However, the recommendations for addressing these concerns focus solely on changing science content and assessment. There is no apparent consideration given to developing understandings of scientific knowledge production and of the socially constructed nature of science, nor to the challenges posed by recognition of its gendered nature and multiculturalism. Neither do the recommendations discuss the roles that normative and monetary power play in the making of science nor encourage involvement in aspects of society which impinge upon the students' quality of life. The new science curriculum described by Reiss, Millar, & Osborne (1999) includes:

- enhancing general "scientific literacy";
- providing a clear statement of aims, "making clear *why* we consider it valuable for all our young people to study science, and *what* we would wish them to gain from the experience" (p.69);
- presenting the curriculum clearly and simply and following the aims;

- presenting scientific knowledge as a number of key "explanatory stories," including case studies of historical and current issues;
- incorporating technology and applications of science;
- developing understanding of some key "ideas about the ways in which reliable knowledge of the natural world has been, and is being, obtained" (p.69);
- encouraging use of a wide variety of teaching methods and approaches; and
- using assessment approaches which encourage teachers to focus on students' abilities to understand and interpret scientific information and discuss controversial issues.

Global Culture and Science Education

In contrast to the above view of a future direction for science education, there is a growing number of science educators who are working from more global perspectives and relating the various critiques of modern western science to new approaches to science education. Some are writing from feminist perspectives, drawing on the critical and cultural studies of science and technology, to argue for a different approach to science education (see, for example, Brickhouse, 1994; A. Gough, 1998; Kenway & Gough, 1998). Others, such as Cobern (1996b) and Kyle (1999), are writing from a multicultural perspective. A different approach again is put forward by Edgar Jenkins (1994), who is concerned with developing public understanding of science and its relationship with school science education. He argues that

> It is not simply that contemporary constructivist approaches to teaching and learning highlight the importance of students' own conceptions of the natural world. More attention needs to be given to ways of teaching and learning science that might be described as culturally sensitive and to relocating western science within the variety of ways of making sense of the natural world (p.604).

Similarly, Derek Hodson (1998) comments that "If science and technology are driven by sociocultural factors, it follows that different societies will define and organise science differently and so produce different science" (p.98). He concludes that there is much to be gained from confronting students in science classes with questions such as whether it is meaningful to recognize a distinctive "African science," "Aboriginal science," or "feminist science": "How much could be changed (aims, values, methods, criteria of validity, content and so on) and it still be considered science?" (Hodson, 1998, p.98).

Both Noel Gough (1998) and William Cobern (1996a) focus on constructivism and science education, with both authors engaging the problem of "language games" in a global cultural context. "Cobern does not consider that separate indigenous sciences exist, but rather, that different worldviews interpret 'reality' in different ways" (Baker, 1996, p.18), a position that is consistent with Cobern's support for constructivism, but I am sure there would be many writers on indigenous science and local knowledge traditions who would disagree with his position that these are merely "different worldviews." Cobern (1996a) argues that "science education research and curriculum development efforts in non-western countries can benefit by adopting a constructivist view of science and science learning" (p.295). Noel Gough (1998) does not see the situation so simply but argues that there are "issues of 'cultural blindness' that may accompany attempts to implement Western science education programs in cross-cultural and/or multicultural settings"; he believes that constructivism is a "new orthodoxy of Western science education" (p.508) which has limited applicability in non-Western cultural contexts. In framing his argument, Gough draws on many of the same sources to which I have referred above (probably because we share a library!).

Exploring Local Knowledge Traditions in Science Pedagogy
Exploring local knowledge traditions as part of the content of a science curriculum is consistent with Peter McLaren's exhortation to "interrogate *the universal already contained in the local* and examine how the ethnic and regional *is already populated by other perspectives and meanings*" (1995, p.25, emphasis in original) and with Sandra Harding's concern to problematize the universality claims of science. For example, Harding (1998) reminds us that "Local features of knowledge systems are not just 'prison houses of belief' but also 'toolboxes' that enable scientific and technological traditions to understand more about their natural and social environments and interact more effectively with them" (p.190). Thus a science curriculum should be including these "toolboxes" in their content, and there are some examples of science educators writing from this perspective.

David Turnbull (1997), for example, argues for reframing science and other local knowledge traditions. He adopts a poststructuralist (and postcolonialist) perspective and argues that "we need to enable disparate knowledge traditions to work together through the creation of a third space in which the social organization of trust can be negotiated" (1997, p.551). Drawing on examples as diverse as the construction of Chartres Cathedral, the Polynesian colonization of the Pacific, and rice farming in Indonesia, he provides evidence for different cultures having different knowledge spaces with different devices for handling the knowledge. He provides some thought-provoking discussion about local knowledge traditions and their futures which are consistent with the arguments developed by Harding (1998, see especially chapters 4 and 10).

Other writings are currently appearing under the banner of indigenous science, multicultural science, or postcolonial science. Many of these papers focus upon the contribution of particular cultural groups to developing understandings of science from the perspective of local knowledge traditions (see, for example, Harding, 1993a; Peat, 1995; Sardar, 1989) and, occasionally, relating this to science education. For example, William Cobern (1996b) discusses the relationship between western science and traditional culture and science education in Africa. David Baker (1996) provides a short introduction to the meaning of indigenous science in an Australian context and its implications for science education. Michael Christie (1991) discusses some of the myths underlying both Western and Australian Aboriginal knowledge production. He also discusses four principles which he believes characterize the Aboriginal elders' contribution to the science curriculum of his school. These principles are (1991, p.29–31):

- a focus on the context of scientific study: subject matter is to be examined and interpreted only as it is found embedded within its context;
- the multiplicity of perspectives: in science, we are out to construct the fullest and clearest picture of the situation we can by integrating the best of our collective knowledge rather than being out to discover the one true picture of reality;
- a methodology which ensures the ongoing negotiation of knowledge production;
- the building of balance into the scientific picture making.

These principles can also make a contribution to the following discussion of moving toward a pedagogy which embraces strong objectivity.

A Pedagogy Toward Strong Objectivity

For some time Sandra Harding (1986 onwards) has argued that knowledge is fundamentally value laden and that we need more effective conceptions of scientific objectivity incorporating antiracist and antisexist values, what she calls "strong objectivity." But what does this mean within the context of a (post) modern science education?

Noel Gough (1998) makes one suggestion which can help frame an analysis of recent curriculum reforms in science education:

> Although we might not be able to speak from outside our own Eurocentrism, continuing to ask questions about the globalization of the cultural practices we call science education will, I hope, help to make both the limits—and strengths—of the knowledge tradition we call Western science increasingly visible (p.522).

Another frame comes from a comparison of the approaches to the subjects/agents of knowledge and of knowledge production in the empiricist epistemology of modern science with the approaches in standpoint epistemology (see Table 1).

Other contributors toward pedagogies of science informed by global perspectives which could move toward strong objectivity include the previously mentioned model for curriculum transformation in science education (Kreinberg & Lewis, 1996), border pedagogy (Giroux, 1991), and Aboriginal science principles (Christie, 1991).

Earlier in this chapter I discussed the Nuffield Foundation's new science curriculum for England and Wales, "Beyond 2000" (Reiss, Millar, & Osborne, 1999) and suggested that its deficiencies include an over-emphasis on content and assessment and its numerous silences on the socially constructed nature of science, scientific knowledge production, the role power plays in the making of science, and the potential for public participation in social action.

In the following discussion I focus upon proposals from Solomon (1999), Aikenhead & Jegede (1999), and Osborne and Barton (1998) and then reexamine an earlier proposal of my own (A. Gough, 1998) with a view to addressing the question, "what would a science education consistent with strong objectivity look like?"

Joan Solomon (1999) discusses a "radical" science curriculum which is concerned with:

- The transmission of a *scientific culture*.
- The development of the *cognitive and evaluative skills* of learners.

- Enabling young people to take part in the *reconstruction of their society* with respect to technical and scientific issues (STS) (p.10, emphasis in original).

She sees this curriculum as being designed to develop "a wide but not necessarily deep knowledge of science and an enthusiasm for it which will breed confidence in using easy and vivid parts of its language...[which] is precisely the kind of understanding which has been shown by research to be essential to enable young people to take part in the discussion of social issues" (1999, p.10). Solomon's proposal may be an improvement on the Nuffield Foundation's proposal (Reiss, Millar & Osborne, 1999), insofar as it engages with science culture and social reconstruction, but it still does not problematize either the universal knowledge claims of the science subject matter to be included or essentialized understandings of student identities. Science culture is envisaged as a transmutation of science into general knowledge without any discussion of its social constructedness (including issues of gender and multiculturalism), and there is no consideration of student differences in relation to gender, race, class, or other cultural variables. There is one small hint of a connection between Harding's and Solomon's suggestions: both refer to the history of science. For Harding (1994) this is "a new 'science education' about the history of scientific traditions" (p.327), whereas for Solomon (1999) one step along the road to a popular scientific culture is to "teach using stories from the history of science to gain some understanding of the tentative and humanist nature of its theories" (p.8).

Aikenhead and Jegede (1999) are concerned with the cultural border crossings and collateral learning that occurs when "students move between their everyday life-world and the world of school science" and define the development of "culturally sensitive curricula and teaching methods that reduce the foreignness felt by students" (p.269) as an emerging priority for the twenty-first century. While on the surface this approach might seem to be in accord with a pedagogy informed by strong objectivity, the detail reveals successful border crossings into the culture of science classrooms and science learning are achieved by changing the student rather than the science: the marginalization of both Western and non-Western students "might be reduced by introducing explicit border crossings in the science classroom and by coaching students to cross those cultural borders with greater ease" (Aikenhead & Jegede, 1999, p.276) and by using collateral learning to probe what occurs in the minds of learners. These approaches may help address the concerns of those interested in developing science-for-all programs, but they do not contribute to a pedagogy of strong objectivity as there is no interest in changing the science.

Osborne and Barton (1998) propose a liberatory pedagogy in science "focused on 'at-risk' students and those underrepresented in the science community by race, class or gender" (p.252). Their work is grounded in critical, feminist and poststructuralist theories and social constructivist philosophy as well as border pedagogy (Giroux, 1991, referred to in Barton, 1997). They also consider the interconnected forms of oppression generated by "contradictions between the unproblematic way in which science is presented and the embeddedness of gendered, raced, and classed values in the process and content of science" (1998, p.255) and decenter "the reform focus from the deficiencies held by women and minorities to deficiencies and discriminatory practices in science and education" (1998, p.256). Their approach (described in more detail in Barton, 1998) is consistent with Giroux's (1991) view of border pedagogy but contrasts with the approach proposed by Aikenhead and Jegede (1999), where the deficiencies are assumed to be in the students.

Osborne & Barton (1998) also pose two important questions which are in accord with a pedagogy toward strong objectivity: "Can a science and science education be constructed that is liberatory, rather than oppressive, to those students who have been marginalized by the science endeavour?" and "Can we teach a science which is open to multiple ways of knowing in order to help *all* students value the contributions made by those traditionally silenced in science?" (p.256). I find resonances between these questions and four guiding principles for science education research that I have proposed elsewhere (A. Gough, 1998, p.196):

- to recognize that knowledge is partial, multiple, and contradictory,
- to draw attention to the racism and gender blindness in science education,
- to develop a willingness to listen to silenced voices and to provide opportunities for them to be heard,
- to develop understandings of the stories of which we are a part and our abilities to deconstruct them.

These principles take into account both feminist and global cultural considerations and while I do not consider them to be unproblematic, they might form the basis for further discussion in science education research and curriculum development. Osborne and Barton (1998) conclude their paper as follows:

> In examining and revisioning concepts of "science," roles, and expectations, students and teachers can move beyond assumptions and beliefs to become empowered and liberated through developing science understandings and by shaping relations of power and privilege in science class to be productive....Thus possibilities for liberatory education exist when students and teachers struggle to understand and critique how their individual and collective social and historical locations shape their relationships with each other, science, education, and society. We believe unresolvable questions remain about the effectiveness and purposes of such teaching (p.259).

While I recognize that there is an element of tactical rhetoric in constructing such equivocal concluding sentences, my background in environmental education leads me to believe that the purposes and effectiveness of such teaching are not "unresolvable" questions. German adults may see science as equating to risk and understand ecology not necessarily as a science "but as a new holistic approach to all aspects of life and nature" and as the savior (Bader, 1993, in Solomon, 1999, p.9). However, ecology is generally accepted as a science, albeit one with socially critical/liberatory potential (see Greenall Gough & Robottom, 1993), and an understanding of it, within the frameworks proposed here, is an important part of achieving a democratic future. If we want all groups in our societies to make intelligent decisions about their own lives and the public policies that affect their lives, then we need science educations to be democratically distributed. As John Dewey argued, "an effective pursuit of democracy requires those who bear the consequences of decisions have a proportionate share in making them" (in Harding, 1993b, p.3).

Conclusion

Whichever way you look at it, a science education grounded in modern science is in trouble: it is more a science education for the nineteenth century than the twenty-first, it has an overemphasis on arbitrarily privileged content, does not relate to society at large, and is of declining interest to students. Many attempts are being made to reform the science curriculum, but most are focused on changing the student rather than the subject matter or on an unproblematized science rather than an open-to-critique science. What is needed is a (post) modern science education, one that is democratic and recognizes the socially constructed, gendered, and multicultural nature of science in its global context. It also needs a pedagogy which draws attention to the processes through which knowledge is produced and recognizes that the agents of knowledge are embodied, visible, and socially located. Sandra Harding's strong objectivity is very relevant to changing science education because it brings epistemological, feminist, and postcolonialist critique together with cultural studies of science. Osborne and Barton (1998) have made a worthy step toward enacting such a vision with their liberatory pedagogy in science, but we need many more—and more compelling—examples. There is no one true story for a (post) modern science education, but if we enact a pedagogy oriented toward strong objectivity we may at least teach less partial and less distorted stories of science.

Endnotes

1. By "Western" I mean not only Europe and North America but anywhere else that follows the canons of Enlightenment science. I recognize that by using "Western" I am reinstating the West-East contrast that postcolonial writers are trying to undermine, but this is the construct in use and readily understood so I will continue to use it with the understanding that it should always be read in "scare" quotation marks.

2. For an overview of these critiques, sometimes referred to as the "Science Wars," see A. Gough, (1998), Harding (1994, 1998), Hess (1997), and Rose (1997).

References

Aikenhead, G. S. & Jegede, O. J. (1999). Cross-cultural science education: A cognitive explanation of a cultural phenomenon. *Journal of Research in Science Teaching* 36(3): pp.269–287.

Alcoff, L. & Potter, E. (1993). Introduction: When feminisms intersect epistemology. In Linda Alcoff & Elizabeth Potter (eds.) *Feminist epistemologies*. New York and London: Routledge, pp.1–14.

Baker, D. (1996). Does "Indigenous Science" really exist? *Australian Science Teachers' Journal* 42(1): pp.18–20.

Baker, D. (1998). Equity issues in science education. In B.J. Fraser & K.G. Tobin (eds.), *International handbook of science education*. Dordrecht: Kluwer, pp.869–895.

Baker, D. & Leary, R. (1995). Letting girls speak out about science. *Journal of Research in Science Teaching* 32(1): pp.3–27.

Barton, A. C. (1997). Teaching science with homeless children: Pedagogy, representation, and identity. Paper presented at the AERA annual meeting, Chicago, IL, March 1997.

Barton, A. C. (1998). Reframing "science for all" through the politics of poverty. *Educational Policy*, 12(5): pp.525–541.

Berman, R. (1989). From Aristotle's dualism to materialist dialectics: feminist transformation of science and society. In Alison M. Jaggar & Susan Bordo (eds.) *Gender/body/knowledge: Feminist reconstructions of being and knowing*. New Brunswick, NJ/ London: Rutgers University Press, pp.224–255.

Bleier, R. (1986). Lab coat: Robe of innocence or Klansman's sheet? In Teresa De Lauretis (ed.) *Feminist studies/Critical studies*. Bloomington: Indiana University, pp.55–66.

Brickhouse, N. (1994). Bringing in the outsiders: Reshaping the sciences of the future. *Journal of Curriculum Studies* 26(4): pp.401–416.

Christie, M. J. (1991) Aboriginal science for the ecologically sustainable future. *Australian Science Teachers' Journal* 37(1): pp.26–31.

Cobern, W. W. (1996a). Constructivism and non-Western science education research. *International Journal of Science Education* 18(3): pp.295–310.

Cobern, W. W. (1996b). Traditional culture and science education in Africa: Merely language games? Paper presented at the meeting for Traditional Culture, Science and Technology, and Development: Toward a New Literacy for Science and Technology, Tokyo Institute of Technology Meguro-ku, Tokyo, Japan, 28 September 1996.

Cohen, L. (1994). Whodunit?—Violence and the myth of fingerprints: Comment on Harding. *Configurations* 2: pp.343–347.

Dekker, J. & De Laeter, J. R. (1997). The changing nature of upper secondary school science subject enrolments. *Australian Science Teachers' Journal* 43(4): pp.35–41.

Driver, R., Leach, J., Millar, R. & Scott, P. (1996). *Young People's Images of Science*. Buckingham: Open University.

Farquhar, J. (1994). Political economies of knowledge: Comment on Harding. *Configurations* 2: pp.331–335.

Gilligan, C. (1982). *In a different voice: Psychological theory and women's development*. Cambridge, MA/ London: Harvard University.

Giroux, H. A. (1991). Postmodernism as border pedagogy: Redefining the boundaries of race and ethnicity. In Henry A. Giroux (ed.), *Postmodernism, feminism, and cultural politics: Redrawing educational boundaries*. Albany, NY: State University of New York Press, pp.217–256.

Gore, J. M. (1993). *The struggle for pedagogies: Critical and feminist discourses as regimes of truth*. New York: Routledge.

Gough, A. (1994) Fathoming the fathers in environmental education: A feminist poststructuralist analysis. Unpublished doctoral dissertation, Faculty of Education, Deakin University, Geelong, Victoria, Australia.

Gough, A. (1997). *Education and the Environment: Policy, trends and the problems of marginalisation.* Melbourne: Australian council for Educational Research.

Gough, A. (1998). Beyond eurocentrism in science education: Promises and problematics from a feminist poststructuralist perspective. In William F. Pinar (ed.), *Curriculum: Toward new identities.* New York/London: Garland, pp.185–209.

Gough, N. (1998). All around the world: Science education, constructivism, and globalization. *Educational Policy* 12(5): pp.507–524.

Greenall Gough, A. & Robottom, I. (1993). Towards a socially critical environmental education: Water quality studies in a coastal school. *Journal of Curriculum Studies* 25(4): pp.301–316.

Haraway, D. J. (1991). *Simians, cyborgs, and women: The reinvention of nature.* London: Free Association Books.

Harding, S. (1986). *The science question in feminism.* Ithaca, NY: Cornell University.

Harding, S. (1991). *Whose science? Whose knowledge? Thinking from women's lives.* Ithaca, NY: Cornell University.

Harding, S. (ed.), (1993a). *The "racial" economy of science: Toward a democratic future.* Bloomington: Indiana University.

Harding, S. (1993b). Introduction: Eurocentric scientific illiteracy—A challenge for the world community. In Sandra Harding (ed.), *The "racial" economy of science: Toward a democratic future.* Bloomington: Indiana University, pp.1–22.

Harding, S. (1993c). Rethinking standpoint epistemology: "What is strong objectivity?" In Linda Alcoff and Elizabeth Potter (eds.), *Feminist epistemologies.* New York/London: Routledge, pp.49–82.

Harding, S. (1994). Is science multicultural? Challenges, resources, opportunities, uncertainties. *Configuration* 2: pp.301–330.

Harding, S. (1998). *Is science multicultural? Postcolonialisms, feminisms, and epistemologies.* Bloomington: Indiana University.

Hess, D. J. (1997). *Science studies: An advanced introduction.* New York/London: New York University.

Hodson, D. (1998). Is this what scientists do? In Jerry Wellington (ed.), *Practical work in school science: Which way now?* London/New York: Routledge, pp.93–108.

Hubbard, R. (1981) The emperor doesn't wear any clothes: the impact of feminism on biology. In Dale Spender (ed.), *Men's studies modified: The impact of feminism on the academic disciplines.* Oxford: Pergamon Press, pp.213–235.

Jenkins, E. W. (1994). Public understanding of science and science education for action. *Journal of Curriculum Studies* 26(6): pp.601–611.

Keller, E. F. (1985). *Reflections on gender and science.* New Haven/London: Yale University.

Keller, E. F. (1992). *Secrets of life, secrets of death: Essays on language, gender and science.* New York and London: Routledge.

Keller, E. F. & Longino, H. E. (eds.) (1996). *Feminism and science.* Oxford: Oxford University.

Kenway, J. & Gough, A. (1998) Gender and science education: a review with 'attitude.' *Studies in Science Education* 31: pp.1–30.

Kreinberg, N. & Lewis, S. (1996). The politics and practice of equity: Experiences from both sides of the Pacific. In L.H. Parker, L.J. Rennie, & B.J. Fraser (eds.), *Gender, science and mathematics: shortening the shadow.* Dordrecht: Kluwer, pp.177–202.

Krockover, G. H. & Shepardson, D. P. (1995). Editorial: The missing links in gender equity research. *Journal of Research in Science Teaching* 32(3): pp.223–224.

Kuriyama, S. (1994). On knowledge and the diversity of cultures: Comment on Harding. *Configurations* 2: pp.337–342.

Kyle, W. C. Jr (1999). Science education in developing countries: Challenging First World hegemony in a global context. *Journal of Research in Science Teaching* 36(3): pp.255–260.

Lather, P. (1991). *Getting smart: Feminist research and pedagogy with/in the postmodern.* New York/London: Routledge.

McLaren, P. (1995). *Critical pedagogy and predatory culture: Oppositional politics in a postmodern era.* Routledge, New York.

Medway, P. (1993). *Shifting relations: Science, technology and technoscience.* Geelong: Deakin University.

Merchant, C. (1980). *The death of nature: Women, ecology and the scientific revolution.* New York: Harper and Row.

Namenwirth, M. (1986). Science seen through a feminist prism. In Ruth Bleier (ed.), *Feminist approaches to science.* New York: Pergamon, pp.18–41.

Osborne, M. & Barton, A. M. C. (1998). Constructing a liberatory pedagogy in science: Dilemmas and contradictions. *Journal of Curriculum Studies* 30(3): pp.251–260.

Peat, F. D. (1995). *Blackfoot physics*. London: Fourth Estate.

Reiss, M., Millar, R. & Osborne, J. (1999). Beyond 2000: Science/biology education for the future. *Journal of Biological Education* 33(2): pp.68–70.

Rose, H. (1997). Science wars: My enemy's enemy is—only perhaps—my friend. In Ralph Levinson & Jeff Thomas (eds.), *Science today: Problem or crisis?* London/New York: Routledge, pp.51–64.

Rosser, S. V. (1997). *Re-Engineering female friendly science*. New York: Teachers College.

Sardar, Z. (1989). *Explorations in Islamic science*. London/New York: Mansell.

Schuster, M., & Van Dyne, S.(1984). Placing women in the liberal arts: Stages of curriculum transformation. *Harvard Educational Review* 54(4): pp.413–430.

Solomon, J. (1997). Girls' science education: Choice, solidarity and culture. *International Journal of Science Education* 19(4): pp.407–417.

Solomon, J. (1999). Meta-scientific criticisms, curriculum innovation, and the propagation of scientific culture. *Journal of Curriculum Studies* 31(1): pp.1–15.

Stanley, L. & Wise, S. (1990). Method, methodology and epistemology in feminist research processes. In Liz Stanley (ed.), *Feminist praxis*. London/New York: Routledge, pp.20–60.

Turnbull, D. (1991). *Technoscience worlds*. Geelong: Deakin University.

Turnbull, D. (1997). Reframing science and other local knowledge traditions. *Futures* 29(6): pp.551–562.

Weedon, C. (1987). *Feminist practice and poststructuralist theory*. Oxford: Blackwell.

Weedon, C. (1999). *Feminism, theory and the politics of difference*. Oxford: Blackwell.

Contributors

Karen Anijar is associate Professor of Education at Arizona State University. She is the author of *Teaching Toward the 24th Century: Star Trek as Social Curriculum*.

Peter Appelbaum is Associate Professor of Mathematics and Science Education at William Paterson University in New Jersey. He is the author of *Popular Culture, Educational Discourse, and Mathematics* and is a contributor to numerous educational journals.

David W. Blades is Associate Professor of Science Education at the University of Alberta. He is the author of *Procedures of Power and Curriculum Change: Foucault and the Quest for Possibilities in Science Education*. He has lectured throughout the world on science education issues.

William E. Doll, Jr. is Vira Franklin and J.R. Eagles Professor of Curriculum at Louisiana State University in Baton Rouge. He is the author of *A Postmodern Perspective on Curriculum*. He has written numerous articles on curriculum theory and postmodernity and has lectured throughout the United States and Canada.

Franc Feng is an assistant professor at British Columbia University in Vancouver, Canada.

Susan Gerofsky is a mathematics educator with a background in linguistics, languages, and literature. She has worked extensively in the field of film production as a picture and sound editor. She is currently completing her doctorate at Simon Fraser University and is a teacher at an alternative high school in Vancouver.

Jeffrey Glanz is an Associate Professor in the Department of Instruction, Curriculum, and Administration at Kean University in New Jersey.

Annette Gough is Senior Lecturer in environmental and science Education in the Faculty of Education, Deakin University, Australia. She is also vice president of the Science Teachers Association of Victoria and a life fellow of the Australian Association for Environmental Education.

Her major areas of research interest include effective teaching and learning in science and the intersection of feminism and environmental education in a research context.

Noel Gough is an Associate Professor of Science Education at Deakin University in Australia. He is the author of *Laboratories in Fiction: Science Education and Popular Media*. Gough also has written widely on science education, the posthuman condition, science fiction, and environmental education.

Elaine V. Howes is Assistant Professor of Science Education at Columbia University.

Eva Krugly-Smolska is an Associate Professor in the Faculty of Education, Queen's University at Kingston and part of the Mathematics, Science, Technology Education group. Her research interests focus on comparative science education and the interrelationships between science, education, and culture.

Marla Morris is a Assistant Professor of curriculum theory at Georgia Southern University. She is the co-editor, with John A. Weaver, of the forthcoming book *Difficult Memories: Talk in a Post-Holocaust Age*, and the author of *Curriculum and the Holocaust: Competing Sites of Memory and Representation*, and co-editor with Mary Aswell-Doll and William Pinar of *How We Work* also with Peter Lang.

Stephen Petrina is an Assistant Professor at British Columbia University, Vancouver, Canada.

David Pushkin is an Assistant Professor of Educational Research at Wilmington College of Delaware. He is the author of numerous chapters on science education, critical pedagogy, and cognitive development.

Bill Rosenthal is Assistant Professor of Mathematics Education at Muhlenberg College in Pennsylvania.

Matthew Weinstein is Associate Professor and Director of Secondary Education at Macalester College in Minnesota. When not supervising student teachers, Matthew conducts research on the intersections of schooling, science, popular culture, and political struggle. He is the author of *Robot World*.

John A. Weaver is Associate Professor of Educational Foundations and Leadership at the University of Akron. He is the co-editor of *Popular Culture and Critical Pedagogy* with Toby Daspit. He is also co-editor with Toby Daspit and Karen Anijar of *SF Curriculum, Cyborg Teachers, and Youth Cultures* (Peter Lang) and author of *Re-thinking Academic Politics in (Re-)Unified Germany and the United States*.

Index

A

Acheson, Keith, 229
Aikenhead, Glen, 71, 77, 292–294
Alcoff, Linda, 278
Aldini, Giovanni, 49
Allegre, Claude, 188
Althusser, Louis, 41
Anaximander, 99
Appelbaum, Peter, 115–116, 171
Apple, Michael, 136
Archer, Mike, 259–260, 265
Aristotle, 42, 104, 106, 183–184, 186
Asimov, Isaac, 257
Atkinson, Paul, 200
Atwater, M.N., 188

B

Babich, Babette, 4, 7
Bacon, Francis, 5, 42
Bachelard, Gaston, 23, 41–54
Bair, Alvin, 222
Baker, David, 281–282, 291
Bakhtin, Mikhail, 2, 15, 147–148
Ballard, J.G., 249–253
Balliet, T.M., 216
Bartolme, L., 203
Bash, Barbara, 116
Bataille, Georges, 3
Bateson, Gregory, 37
Baudrillard, Jean, 1, 23–24, 57, 59, 61–66, 68–69, 72, 74–74, 78–86, 88–90, 123, 249, 252–253
Beardsley, Tim, 71
Belenky, Mary, 115
Bergmann, M., 170
Berman, Morris, 37
Berman, Ruth, 279
Bernstein, Richard, 37
Bérubé, Michael, 15
Bierhorst, John, 60
Bishop, Alan, 179
Blades, David, 23–24, 69–71, 76–77, 81
Blake, William, 267
Blanchot, Maurice, 95
Bleier, Ruth, 280
Bley, Michel, 189
Bloor, David, 134
Bobbitt, Franklin, 217–220
Bohr, Neils, 10–13
Bolin, F. 217
Boll, Marcel, 54
Boomer, Garth, 258
Boorstin, Daniel, 73
Boulter, C.J., 203
Boundas, Constantin, 139
Bouty, Edmond, 53
Bowers, Chet, 230
Bowers, J.W., 170
Boyce, Arthur, 221
Boyd, William, 256
Boyle, Robert, 5
Brickhouse, Nancy, 280, 289
Bricmont, Jean, 5, 7–8, 111
Brillouin, Léon, 96, 105
Brogan, Nancy, 233, 235
Bruchac, Joseph, 86
Brueckner, L.J. 224
Bukatman, Scott, 249–250
Burroughs, William, 251

Buscombe, Edward, 147
Bybee, Rodger, 79

C

Cabrese-Barton, Angela, 278, 281, 294–296
Caduto, Michael, 91
Cantor, Georg, 185–186
Carlsen, William, 134
Casey, Nancy, 187
Cherryholmes, Cleo, v, 62
Chiang, C-L., 90
Christie, Michael, 291
Clark, Arthur, C., 254
Clark, Stella, 115, 171
Clinton, Bill, 131
Cobern, William, 289–291
Coffey, Amanda, 200
Cohen, Jack, 14–15, 260, 264, 266
Cohen, Lawrence, 284
Coleridge, Samuel, 267–268
Collins, Harry, 112–113
Colón-González, M.H., 194, 209
Cooper, Andrew, 269
Coppola, B.P., 208
Cordeiro, Pat, 187
Cordelio, Pat, 74
Costa, Arthur, 228
Courtis, S.A., 218–219
Crawford, T. Hugh, 17–19
Csicsery-Ronay, Istvan, 250, 252, 254
Cunningham Christine, 134

D

Dalthorp, Charles, 220
Damarin, Suzanne, 189
D' Aragona, Julia, 177
Daston, Lorraine, 140
Dauben, Joseph, 185–186
Davidson, William, 220
Davis-Floyd, Robbie, 136
Dekker, J., 287
DeLaeter, J.R., 287
Delaney, Samuel, 252
Derrida, Jacques, 6, 41, 60–62, 243–244
Desalle, Rob, 260–261
Descartes, Rene, 104–105
Dewey, John, 69, 107, 115, 226–227, 295
Dickens, Charles, 200
Dinnerstein, Dorothy, 180
Dirac, Paul, 46
Disney, Walt, 77
Doll, William, 23
Dolly, the Sheep, 64
Donne, John, 254
Driscoll, Marcy, 197
Dubeck, Leroy, 263
Duit, Reinders, 203
Dumas, Alexander, 26
Duschl, Richard, 77
Dyson, Freeman, 245–246

E

Edwards, Richard, 12
Einstein, Albert, 46, 142
Eisler, Riane, 213
Eisner, Elliot, 213
Eisenhart, Margaret, 142
Eliot, Edward, 221
Elsbree, Willard, 215, 224
Ervin-Tripp, Susan, 164, 166
Epstein, Steven, 136

Index

F

Fairclough, Norman, 250
Falk, R., 189
Farquhar, Judith, 284
Feigenbaum, Mitchell, 14
Feng, Franc, 23
Ferguson, Charles, 169
Finkel, Elizabeth, 141–142
Fiore, Quintin, 257
Fitzpatrick, F.A., 217
Flanagan, Martin, 267, 270
Fletcher, M.A., 130
Flinders, David, 230
Fossey, Dian, 256
Foster, Henry, 130
Foucault, Michel, 6, 41, 63, 250
Freire, Paulo, 71, 201, 204

G

Gabriel, Joseph, 138
Gadamer, Hans-Georg, 98
Galileo, Galilei, 18, 44
Galison, Peter, 135, 245
Gall, Meredith, 229
Gallas, Karen, 119
Garman, Noreen, 213
Garmston, Robert, 228
Geduld, Harry, 256
Gerofsky, Susan, 122, 153
Gibson, William, 251
Gilbert, J.K., 203
Gilland, Thomas, 215
Gilligan, Carol, 282
Giroux, Henry, 285, 292, 294
Glanz, Jeffrey, 123, 214–215, 232–233
Glickman, Carl, 213–214, 228–229

Gluscabi, the hunter, 86–87
Goffman, Erving, 161, 163
Goldhammer, Arthur, 52
Goldhammer, Robert, 228
Goodall, Jane, 256
Gordon, Beverly, 207
Gordon, Stephen, 213–214, 228–229
Gore, Jennifer, 278
Gough, Annette, 247, 276–278, 289, 292, 294–295
Gough, Noel, 75–76, 114, 124, 246–247, 252–254, 258, 290–291
Greenwood, James, 215–216
Greenwood, Thomas, 183
Griffin, Susan, 180
Griffiths, Daniel, 215
Grills, Scott, 200
Grindall, Karen, 118
Gross, Alan, 16
Gross, Paul, 1–4, 111, 141, 246
Grumet, Madeleine, 215
Gunstone, Richard, 198
Guo, C.J., 90
Guy, R. 67–68
Gwynn, John, 215

H

Habermas, Jürgen, 30
Hacking, Ian, 133–134
Hammer, D. 199
Hammersley, Martin, 200
Haraway, Donna, 5–6, 17, 111, 133–134, 136–137, 140, 247, 249, 253–258, 276, 283
Harding, Sandra, 247, 275–280, 282–285, 287, 290–291, 293, 295–296

Harris, James., 130
Harris, Mary, 189
Harris, Michael, 68
Have, ten, Paul, 200
Hayles, N. Katherine, v, 13–14, 19, 254, 264
Heidegger, Martin, 23, 27–29, 36–37
Heisenberg, Werner, 10–11, 13
Heraclitus, 99–100
Hermelink, H., 157
Hervey, Henry, 220
Hess, David, 136, 296
Hilbert, David, 177
Hill, Sallie, 224
Hoag, Peter, 256
Hochberg, J.E., 62
Hodson, Derek, 77, 80, 289–290
Hölderlin, Friedrich, 36
Holland, John, 9–10
Holland Patricia, 213
Holt, J. 189
Homer, D., 254
Hopkins, W.S., 230
Horrocks, Chris, 61
Howes, Elaine, 122, 189
Høyrup, Jens, 157-158
Hubbard, Ruth, 279–280
Hurd, Paul, 80, 140
Hyppolite, Jean, 243

J

Jakobson, Roman, 244
James, William, 88
Jamieson, Karen, 150
Jegede, Olugbemiro, 292–294
Jenkins, Edgar, 289
Jeutic, Zoran, 61
Johnson, Franklin, 223

Johnson, Whitney, 181, 189
Jones, James, 130
Jones, Ricky, Lee, 104
Joseph, George, 189
Jourdain, Monsieur, 148
Judd, Charles, 222

K

Kafka, Franz, 154–155
Kass, H., 76
Kaufmann, Walter, 99–101
Keitel, Christine, 155
Keller, Evelyn 139, 262, 266, 279, 283
Keller, Reuben, v
Kellert, Stephen, 13
Kelly, Gregory, 134
Kennedy, Robert, 90
Kenway, Jane, 289
Kess, Joseph, 162–163, 170
Kierrkegaard, Søren, 154
Kincheloe, Joe, 115, 194, 201
King, Leroy, 224
King, Martin Luther, 90
Kirk, Geoffrey, 181
Kitcher, Philip, 59
Koertge, Noretta, 1, 111–114, 117, 119–120
Koshland, Daniel, 266
Kreinberg, Nancy, 281, 292
Krickover, Gerald, 282
Krishnamurti, 79
Krugly-Smolska, Eva, 23
Kuhn, Thomas, 3, 18–19, 133–134, 141, 245
Kuriyama, Shigehisa, 283–284
Kyle, William, 281, 285, 289

Index

L

Lacan, Jacques, 47
Lakatos, Imre, 185
Lakoff, Robin, 170–171
Landow, George, 15
Laplace, Pierre, 13
Laron, Gary, 152–153
Lasley, T.J., 214
Lather, Patti, 4, 283
Latour, Bruno, 1, 6, 8–9, 18–19, 23, 25–38, 96, 111, 133, 135, 247, 254
Lavine, S., 189
Lear, Jonathan, 181, 183, 185
Leary, D., 282
Leavis, Sue, 281, 292
LeGuin, Ursula, 251, 254
Leibniz, Gottfried, 16, 98, 100
Leigh, Julia, 268
Leith, Dick, 163
Lerner, Gerda, 180, 183
Lessing, Doris, 251
Letterman, David, 152
Leunig, Michael, 267
Levin, David, 70
Levy-Bruhl, Lucien, 35
Lewis, Martin, 111
Lewontin, R.C., 266
Lindley, David, 260-261
Lipton, Martin, 124
Longino, Helen, 279
Loving, Cathleen, 1
Lugones, M., 139

M

MacDonald, John, 74
Macedo, Donald, 201
Madison, Gary, 60
Malcolm, Ian, 262
Mandelbrot, Benoit, 19
Maor, Eli, 189
Marion, Scott, 142
Martin, Emily, 133, 136, 138
Maturana, Humberto, 9
Mau, T., 79
May, W.T., 229–231
Maziarz, Edward, 183
McAllester-Jones, Mary, 41, 51–52
McLuhan, Eric, 173
McLuhan, Marshall, 173, 254, 257
McHale, Brian, 250–251
McLaren, Peter, 290
Medway, P., 286–287
Mendel, Gregor, 18
Merchant, Carolyn, 278–279, 287
Merleau-Ponty, Maurice, 95, 101
Merrill, Judith, 249
Merton, Robert, 114, 120
Messer-Davidow, Ellen, 16
Messinger, L., 232
Millar, Robin, 75, 286, 288, 292–293
Miller, Carolyn, 151
Mills, Heidi, 187
Moliere, Jean, 148
Moll, Luis, 197
Moore, A.W., 186, 189
Moore, K.D., 230
Morris, Marla, 24
Murphy, Michael, 269
Myerson, George, 163

N

Namenworth, Marion, 275
Naserin, Farouki, 234
Nehlen, Don, v
Newton, Isaac, 16
Noble, David, 180, 184
Norris, Christopher, 66, 82
Nunan, E.E., 254
Nye, Bill, 77–78

O

Oakes, Jeannie, 124
O'Connor, Flannery, 154
O'Keefe, Timothy, 187
Ollerenshaw, J.A., 200
Osborn, J. 286, 288, 292–293
Osborn, Margery, 170, 281, 294–296

P

Paerson, W.H., 208
Pagano, Lynn, 233, 235
Paino, J., 232
Pajak, Edward, 217
Palmer, Jerry, 147
Panaritis, P., 217
Parson, Talcott, 114, 120
Patterson, James, 136
Paulson, William, 97–98, 106
Payne, William, 215
Peat, F. David, 291
Peirce, C.S., 244
Penick, John, 76
Penley, Constance, 140
Percival, Ian, 14
Perry, William, 194, 196–197, 201, 204

Petrina, Stephen, 23
Pickover, Clifford, 189
Piercy, Marge, 251
Pimm, David, 166, 168–169
Pinar, Bill, 200
Pinch, Trevor, 112–113
Plato, 100, 182, 186
Plotnitsky, Arkady, 2, 12–13
Pohly, Kenneth, 233
Pollard, Charles, 136
Porritt, Jonathan, 267
Potter, Elizabeth, 278
Powell, Isaac, 129
Prigogine, Ilya, 9
Pulley, Jerry, 214
Pushkin, Dave, 123, 193–199, 201, 204, 209
Pynchon, Thomas, 251, 259
Pythagoras, 99-100, 180

Q

Quinn, Daniel, 256

R

Ramses, II, 64
Rapp, Royna, 136
Raven, John, 181
Reinsch, N.L., jr., 170
Reiss, M., 286, 288, 292–293
Reller, Theodore, 215
Ricoeur, Paul, 98
Ring, Merrill, 181
Robbins, Bruce, 246
Robottom, I., 295
Rodriquez, Alberto, 131, 142
Rose, Hilary, 296
Ross, Andrew, 17, 111–113, 140, 245–246

Ross, Dorothy, 16
Ross-Gordon, J.M., 228–229
Ross-Lee, Barbara, 130
Rosser, Sue, 281–282
Rosenthal, Bill, 122, 181, 187
Rotman, Brian, 184
Rouse, Joseph, 133
Rucker, Rudy, 182
Rugg, Harold, 223

S

Sagan, Carl, 78, 88
Sandell, Rolf, 170
Santayana, Georges, 59
Sardar, Ziauddia, 291
Saussure, Ferdinand, 244
Scheurich, James, 200
Schieninger, Linda, 184
Schimmel, Annemarie, 182
Schleiermacher, Friedrich, 97
Schmidt, Rosemarie, 162–163, 170
Scholes, Robert, 9
Schopenhauer, Arthur, 217
Schuster, M., 281
Seale, Clive, 200
Sergiovanni, Thomas, 213–214, 217
Serres, Michel, 1, 24, 38, 95–96, 99–108, 111
Shalala, Donna, 130
Shannon, Claude, 96–97
Shaw, Hermann, 131
Shepardson, Daniel, 282
Simplicius, 99
Singmaster, David, 157
Slothrop, Tyrone, 259
Smyth, John, 213
Snow, C.P., 106

Sobchack, Vivian, 147
Sokol, Alan, 1, 5, 7–8, 111, 243–246
Soloman, Joan, 280, 282–283, 287–288, 292, 295
Souque, Joseph, 47
Spielberg, Steven, 260
Staal, J.F., 73
Stacey, Jackie, 136
Stanley, Liz, 278
Starratt, R.J., 214
Staver, J.R., 194
Stengers, Isabella, 23, 34–35
Stepenson, Neil, 251
Stewart, Ian, 14–15, 260, 264, 266
Stone, Merlin, 180
Stone, T.W., 218–219
Strachota, Bob, 120
Strage, A., 164, 166
Strugatsky, Arkady, 251
Strugatsky, Boris, 251
Stump, D.J., 135
Suchman, Lucy, 137
Sullivan, Susan, 233
Susskind, Y., 141
Swales, John, 151
Sweeney, Leo, 181

T

Tarkovsky, Andrei, 251
Taylor, Elizabeth, 249
Taylor, Frederick, 217–218
Taylor, Joseph, 228
Thales, 99
Thomas, R., 185
Tobias, Sheila, 203
Tomizuka, Carl, 203

Traweek, Sharon, 5, 20, 133, 136, 138, 140
Treagust, David, 203
Treichler, Paula, 136
Tudor, Andrew, 148–149
Turnbull, David, 277, 290
Tzu, Sun, 158

U

Usher, Robin, 12

V

Valdés, G., 200
Van Dyne, S., 281
Varela, Francisco, 9
Varley, John, 255–256
Vilenkin, Naum, 189
Vygotsky, Lev, 197–198, 201

W

Waite, Duncan, 213, 228, 230
Waterson, G.C., 52
Watson, James, 138, 262
Weaver, John, A., 100, 107, 118
Weaver, Warren, 96–97
Weeden, Chris, 283
Weinberg, Steven, 138, 243–245
Weinstein, Matthew, 114, 120, 140
Wells, H.G., 256
Wertheim, Margaret, 184
Westerman, Mike, 259–260
White, Rea, 269
White, Richard, 198
Whitin, David, 187
Williams, Antony, 233

Wills, Christopher, 266
Wittgenstein, Ludwig, 217
Wize, Sue, 278
Wong, D., 203
Wong, Mannor, 234
Woodford, James, 259
Woolcott, Alexander, 169
Woolgar, Steve, 135

X

Xenophanes, 101

Y

Yager, Robert, 76
Yager, S.O., 76

Z

Zeidler, Dana, 77
Zeno, 100, 180
Zola, Emile, 254

Studies in the Postmodern Theory of Education

General Editors
Joe L. Kincheloe & Shirley R. Steinberg

Counterpoints publishes the most compelling and imaginative books being written in education today. Grounded on the theoretical advances in criticalism, feminism, and postmodernism in the last two decades of the twentieth century, Counterpoints engages the meaning of these innovations in various forms of educational expression. Committed to the proposition that theoretical literature should be accessible to a variety of audiences, the series insists that its authors avoid esoteric and jargonistic languages that transform educational scholarship into an elite discourse for the initiated. Scholarly work matters only to the degree it affects consciousness and practice at multiple sites. Counterpoints' editorial policy is based on these principles and the ability of scholars to break new ground, to open new conversations, to go where educators have never gone before.

For additional information about this series or for the submission of manuscripts, please contact:
        Joe L. Kincheloe & Shirley R. Steinberg
        c/o Peter Lang Publishing, Inc.
        275 Seventh Avenue, 28th floor
        New York, New York 10001

To order other books in this series, please contact our Customer Service Department:
        (800) 770-LANG (within the U.S.)
        (212) 647-7706 (outside the U.S.)
        (212) 647-7707 FAX

Or browse online by series:
        www.peterlangusa.com